Earth Treasures

Earth Treasures

VOLUME 1: THE NORTHEASTERN QUADRANT

Connecticut, Delaware, Illinois, Indiana, Maine, Maryland, Massachusetts, Michigan, New Hampshire, New Jersey, New York, Ohio, Pennsylvania, Rhode Island, Vermont, and Wisconsin

VOLUME 2: THE SOUTHEASTERN QUADRANT

Alabama, Florida, Georgia, Kentucky, Mississippi, North Carolina, South Carolina, Tennessee, Virginia, and West Virginia

VOLUME 3: THE NORTHWESTERN QUANDRANT

Idaho, Iowa, Kansas, Minnesota, Missouri, Montana, Nebraska, North Dakota, Oregon, South Dakota, Washington, and Wyoming

VOLUME 4: THE SOUTHWESTERN QUADRANT

Arizona, Arkansas, California, Colorado, Louisiana, Nevada, New Mexico, Oklahoma, Texas, and Utah

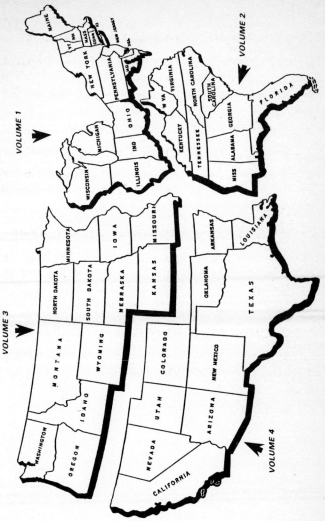

Also by Allan W. Eckert

Bayou Backwaters
Blue Jacket
The Conquerors
The Court-Martial of Daniel
  Boone
The Crossbreed
The Dark Green Tunnel
The Dreaming Tree
The Frontiersmen
Gateway to Empire
The Great Auk
The HAB Theory
In Search of a Whale
Incident at Hawk's Hill
Johnny Logan

The King Snake
The Owls of North America
Savage Journey
The Scarlet Mansion
The Silent Sky
Song of the Wild
Tecumseh!
A Time of Terror
The Wading Birds of North
  America
The Wand
Whattizzit?
Wild Season
Wilderness Empire
The Wilderness War

EARTH TREASURES

VOLUME

The Northwestern Quadrant

Idaho, Iowa, Kansas, Minnesota,
Missouri, Montana, Nebraska, North Dakota,
Oregon, South Dakota, Washington, and Wyoming

by Allan W. Eckert

PERENNIAL LIBRARY

Harper & Row, Publishers, New York
Cambridge, Philadelphia, San Francisco, Washington
London, Mexico City, São Paulo, Singapore, Sydney

To one of the most knowledgeable and
helpful rockhounds I ever met,

William Harris,

of the Harris Rock Shop,
Sarasota, Florida

FIRST EDITION

Designer: C. Linda Dingler

Library of Congress Cataloging-in-Publication Data
Eckert, Allan W.
  Earth treasures.
  "Perennial Library."
  Includes indexes.
  Contents: v. 1. The Northeastern quadrant. v. 2. The Southeastern
quadrant. v. 3. The Northwestern quadrant. v. 4. The Southwestern
quadrant.
  1. Mineralogy—Collectors and collecting—United States. 2. Rocks—
Collectors and collecting—United States. 3. Paleontology—Collectors
and collecting.
I. Title.
QE375.E27   1987   552'.0075'0973   86-45654
ISBN 0-06-096101-5 (pbk. : v. 1)
ISBN 0-06-096131-7 (pbk. : v. 2)
ISBN 0-06-096177-5 (pbk. : v. 3)
ISBN 0-06-096178-3 (pbk. : v. 4)

87 88 89 90 91 MPC 10 9 8 7 6 5 4 3 2 1

CONTENTS

HOW TO USE THIS BOOK

Earth Treasures is geared toward everyone interested in collecting rocks, minerals, and fossils—not only people who are devoted collectors and willing to spend hours, days, sometimes even whole vacations in the pursuit but also those who have only a mild interest and want results quickly without a great deal of wasted effort. Whether these specimens are for personal collections, for trade or sale, for scientific study, for display as cabinet specimens, or for tumbling or cutting into faceted gemstones or cabochons as jewelry makes no difference. It is the purpose of these volumes, through very precise directions and accurate county maps, to get you to the best and most accessible sites as quickly and easily as possible. Each of the four volumes covers one quadrant of the 48 contiguous states, and each volume is an entity in itself, complementing the others but not dependent on any of them.

This particular volume contains directions to 1677 collecting sites and has 558 maps. Each collecting site included is indicated by number on the map for the county in which it is located. This site number also appears under the given county in the text, followed by specific written directions for getting to the site in question and a description of the type (and often the quality) of the rock, mineral, or fossil material to be found there, plus exactly where to search for the material *at the site* (in soil, matrix, road cut, pegmatite, shoreline, quarry, gravel pit, prospect, mine, etc.). Let's take, as an example, under the heading of Nebraska, on the map of Sioux County, the site indicated as number 7. Turning to the proper text page, the corresponding entry is as follows:

7.   Orella Station area (a former watering stop of the Chicago, Burlington & Quincy RR located 20.5 miles north of Crawford in adjoining Dawes County and 5 miles south of Ardmore in Fall River County, South Dakota): 0.5 mile south on SR-29 to right turn on faint dirt road; cross tracks and go 3 miles west through Badlands area (negotiable for autos but four-wheel drive is recommended) to a region of canyons; collect in canyon walls and below them.

Agate, fortification (Fairburn var.). Fossils, marine: some gastropods and corals; some silicification. Fossils, vertebrate: dinosaur and mammalian bone; often well silicified; mammalian bones and frequent teeth. Jasper: red, yellow. Petrified wood. Opal, wood: some of excellent quality. (1-5-sandstone 8-13-14-39)

The initial number, 7, corresponds to the number on the correct county map; in the directions, the first portion—*Orella Station area*—initially directs the reader to the closest municipality of any consequence within that county; in this case, since Orella Station is not shown on many highway maps, extra guidelines are given: the city of Crawford in Dawes County and the city of Ardmore in neighboring South Dakota's Fall River County. The second portion of the entry gives distance and direction—0.5 mile south on State Route 29 to where a dim dirt road intersects from the west; the collector then follows that dim road westward for 3 miles through an area of badlands that can be traveled by auto, although a four-wheel-drive vehicle is safer, to where some canyons are located; the next direction zeroes in on the specific area—in those canyon walls and beneath them. In the next section, devoted to *what* you will find at the site and *where* to look, we see that there are both mineral and fossil deposits—fortification agate of the prized Fairburn variety, two colors of jasper, and wood opal, some of which is excellent; at the same site can also be found well-silicified dinosaur bones as well as mammalian bones and teeth. Finally, the numerals in parentheses following the entry correspond to the mineral location key (found on pages xviii–xix and also following the index) indicating that the specimens are found (1) on the surface and/or in the soil, (5-sandstone) in a matrix of sandstone, (8) in the talus or natural debris at the base of the canyon walls, (13) in washes, ravines, gullies, draws, and ditches, (14) weathered from matrix, and (39) in a natural bed. More expansively, this means that the three minerals and fossils in question, originating from a sandstone matrix material in the canyon walls, are found not only in that matrix, from which they can be dug, but that they have weathered from it into erosion areas, from which they can be collected, and that they are also deposited in the accumulated talus at the base of those canyon walls. Many site directions in this volume contain much less detail than this one; others provide considerably more.

Unless you begin finding considerable worthwhile material in an area where you are searching, it is usually somewhat of a waste of time to dig haphazardly. Wherever possible, put someone else's efforts to your own advantage; it is wise to benefit from the diggings done by others, such as in quarries, gravel pits, sand and clay pits, mine diggings and dumps, abandoned prospects, and excavations being done for highway, canal, railroad, and building construction.

By far the greater majority of the sites indicated here are reachable by auto, and many require very little strenuous hiking after you've parked. Some are reachable by Jeep trail (some for which four-wheel drive is recommended, others for which it is a prerequisite), and a few may demand more extensive walking. Every effort has been made to provide accurate directions, but when following them it should be borne in mind that conditions change constantly: route numbers may be revised or new roads may have been constructed that provide easier access than indicated here; similarly, old roads may have become closed or abandoned or, if unmaintained, may have fallen into impassable condition; some cannot be negotiated during inclement weather conditions; some collecting areas become closed, while others are opened; many areas require permission for entry, and some may charge a small fee or have restrictions concerning the type or amount of material that may be taken. It is your responsibility always to check for local regulations and possible restrictions and to acquire the proper permission if this is necessary.

For those who will be using this guide to stop at numerous locations for collecting (such as during vacation travels), it is wise to carry a supply of plastic bags (the heavy-gauge zip-lock type are especially useful) and a packet of self-adhesive labels. It is, unfortunately, not uncommon for a collector to pick up a selection of specimens from various sites visited and then later (perhaps many months later), when looking them over, discover that he has some especially fine material and no good recollection of exactly where he found those particular specimens. An acquaintance of the author turned out, to his dismay, to be a case in point. On a summer vacation drive through the western states he collected a great many specimens from various stops. Not until the following winter, when he was going through these collected materials and cutting some of them, did he realize that one chunk of quartz he had picked up had a thin vein of native gold running through it. He has no idea where he picked up that particular rock, not even the state. Variations on this theme have occurred to many collectors in the past. Therefore, when *you* are collecting specimens, don't mix them up in a sack or pouch or box containing specimens from several locations. Instead, place all specimens from one location in one bag and then stick a label on the bag and mark it with this guide's page number, location, and date. Thus a bag of specimens labeled 5/21/87—164/6 would later refresh your memory that the specimens in that particular bag were found on May 21, 1987, in the location marked as number 6 on the map on page 164 of this guide.

Bear in mind that many collectors, with justification, carefully guard their secret collecting locations and, if they speak of them at all, are reluctant to reveal their whereabouts except in very nebulous terms. As a result, occasionally some (fortunately, not too many) of the entries in this book will be found to be annoyingly vague. This is in no way meant to be a deliberate effort to mislead; it indicates only that the specimens seen or reported from the location are so good that it was deemed worthwhile to include the information, however limited, the directions provided herein to the site being the best available at the time this volume was prepared. In this same vein, there may also be an occasional site listed that is in error, since in some cases the information has been gathered by word of mouth from collectors and the author may not have been to the site himself. If there are such errors, letters of outrage may be sent to the author in care of the publisher of this book, and corrections will be made in subsequent printings. It is, after all, one of the aims of *Earth Treasures* to make it not only the most comprehensive guide available but also the most *accurate*. Furthermore, the author would appreciate hearing from any users of this book who have collecting sites of their own that they would like to have included in future printings. Such letters should contain all the pertinent information to the best of your ability, with *accurate* directions to the site. Address letters to the author, care of Harper & Row, Publishers, Inc., 10 East 53rd Street, New York, NY 10022.

The purpose of this book and its three companion volumes is solely as a guide to collecting sites. A relatively comprehensive glossary of terms involved in rock, mineral, and fossil collecting will be found at the end of each volume. However, no information is provided as to identifying specific rocks, minerals, or fossils; there are texts available that do this remarkably well, providing keys to identification, hardness scales, specific-gravity charts, photo identification of specimens, and crystal formation information. It would behoove the collector to acquire such books as are needed for identification in the field or at home. Highly recommended as two of the best for use on collecting trips are *The Audubon Society Field Guide to North American Rocks and Minerals* by Charles W. Chesterman (Knopf, 1978) and *Guide to Rocks & Minerals* by Martin Prinz, George Harlow, and Joseph Peters of the American Museum of Natural History (Simon & Schuster, 1978). Two excellent hard-bound at-home books are *Gemstones of North America* and *Prospecting for Gemstones and Minerals*, both by John Sinkankas (Van Nostrand Reinhold, 1959 and 1970). The latter book is undoubtedly one of the better works available on the most successful techniques of collecting.

The county maps in this volume do not include every county of the state in question; only those counties with significant mineral, rock, and fossil collecting sites are included. Additional counties and their maps may be included in future editions. Each of the maps herein, unless otherwise indicated, is oriented with north at the top of the page. In some cases it has been necessary to divide a single county map into two or more portions; San Bernardino County in California, for example, is more than four times larger than the entire *state* of Connecticut.

Do not make the mistake of checking out only the precisely indicated site when visiting any of these locations. Quite often the guide will lead you to collecting areas where, with some diligent searching in adjacent or nearby areas, you may make some noteworthy mineral, rock, or fossil discoveries on your own that have previously lain undetected. In the event that you make such a discovery, there is ample room then to mark in your own site on the map covering that area in this volume. Consistent marking of this guide in that way will, over the years, provide you with a decidedly valuable record.

It should be stressed that because these sites are listed here does not mean that you automatically have the right to collect there. Unless the sites are on public lands (and even some of these require getting a collecting permit), all these sites are on land that is privately owned, and it is your responsibility to ask permission to collect. Failure to do so could result in a confrontation with an angry landowner or possibly even your arrest for trespassing. Remember, too, that collecting rocks, minerals, or fossils is not permitted in national parks and in many state parks. There may also be stringent regulations with respect to collecting on Indian reservations. It is up to you to learn and obey whatever local regulations exist. No matter where you collect, respect the land and anything on it and follow a few simple rules: ask permission, do not litter, be sure to refill any holes you dig, close any gate you open, do not cut or break down fences, do not tamper with or move any equipment, do not molest livestock.

The author, when on collecting expeditions, always carries a stock of release forms to offer to the landowner, quarry operator, mine manager or other person of authority at the collecting site. Signed by you in the presence of this person and given to him, this form will often allay whatever fear may exist on his part that a lawsuit will be brought against him if you are injured while on his property. It also assures him that you are a responsible person who will respect his property, equipment, or livestock. An example of the form will be found immediately following. Having a quantity of such forms on hand will open many potentially good collecting areas for you, as they have for me, to which you might otherwise be denied access. This is especially true in the case of commercial operations, such as mines, quarries, and gravel pits. (Bear in mind as well that most such commercial operations, if they will let you in at all, require that you wear a hardhat—sometimes safety shoes and safety glasses as well—and may refuse entry to you if you do not have such protection.)

RELEASE AND AGREEMENT FORM

[Sample]

————————————,19——

    By signing this paper, I hereby release the owner, proprietor, supervisor, or other official of:

_____

_____

_____

of any responsibility, liability, or claim for damages while I am on the above property. Any personal injury or damage to my property that I may suffer, from any cause whatever, while on this property is deemed by me, the undersigned, as my own responsibility.

    My sole purpose in entering this property is to look for fossils and/or rock and mineral specimens. I agree to respect the property at all times, to stay away from machinery and out of danger zones or places where I might interfere with any operations. I also agree not to park my vehicle where it will block machinery or traffic. Any gate opened by me will be immediately closed again. I will not, in any way, molest any equipment, material, or livestock. In all respects, I agree to treat this property with care and thoughtfulness.

    I understand that my access to this facility is limited to the date on which I sign this release and agreement.

(signed)   _____

(address)   _____

_____

COMPASS DIRECTIONS

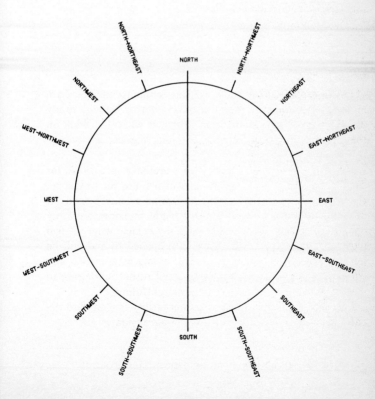

KEY TO PRINCIPAL ABBREVIATIONS AND MINERAL LOCATION NUMBERS

Principal Abbreviations

aka	also known as
BLM	Bureau of Land Management
Byp	bypass route
CF	collecting fee
cm	centimeters
CR	county road
ct	carat
I	interstate highway
lb	pounds
MM	mile marker
mm	millimeters
oz	ounces
RR	railroad
r/w	right-of-way
SR	state highway
US	U.S. highway
var.	variety
"	inches
'	feet

Mineral Location Numbers

1. On the surface and/or in the soil
2. In mine and/or in mine dump
3. In gravel deposit
4. In natural or man-made exposure and/or cut
5. In matrix or *in situ*

6. In soil (but usually *not* on surface)
7. In stream bed, lake bed, coastal shallows
8. In talus and/or natural debris
9. In vugs and/or pockets and/or cavities
10. In gravel pit
11. In sand pit
12. In quarry
13. In washes, ravines, gullies, draws, ditches
14. Weathered from matrix
15. In volcanic tuff or pillow basalt
16. In pegmatite dike
17. In gold-bearing sands and/or gravels
18. In/on gravel bar
19. In vein or seam
20. Loose on shoreline
21. In prospect or placer
22. In sediment deposits or sedimentary strata
23. In alluvial deposits
24. In exposed pegmatite dike
25. In clay
26. In sand
27. In schist
28. In shale and/or slate
29. In outcropping (natural or exposed by man)
30. In spoil deposits
31. In ash deposits
32. In cave and/or cave vicinity
33. In marl
34. In gneiss
35. In clay pit
36. In/on sand bar
37. In/on fill deposit
38. Occurring as drusy or crust
39. In natural bed or deposit
40. In conglomerates
41. In concretions
42. In/on ledges

IDAHO

Statewide total of 82 locations in the following counties:

Ada (1)

Adams (4)

Bannock (1)

Bear Lake (1)

Benewah (3)

Bingham (1)

Blaine (4)

Boise (3)

Bonner (1)

Boundary (1)

Butte (1)

Canyon (3)

Caribou (1)

Cassia (1)

Clearwater (1)

Custer (6)

Elmore (5)

Fremont (1)

Gem (2)

Gooding (1)

Idaho (5)

Kootenai (2)

Latah (2)

Lemhi (7)

Lewis (1)

Lincoln (1)

Nez Perce (2)

Owyhee (6)

Shoshone (5)

Twin Falls (2)

Valley (3)

Washington (4)

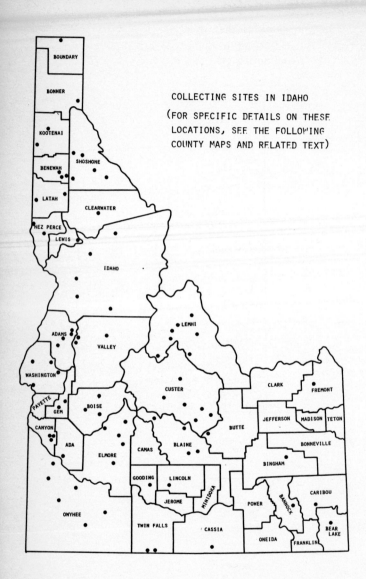

COLLECTING SITES IN IDAHO

(FOR SPECIFIC DETAILS ON THESE
LOCATIONS, SEE THE FOLLOWING
COUNTY MAPS AND RELATED TEXT)

ADA COUNTY

1. Swan Falls area: on the east shore of the Snake River.
   Diamond: 1 stone having unconfirmed weight of 2.125 ct; found during gold panning. (17)

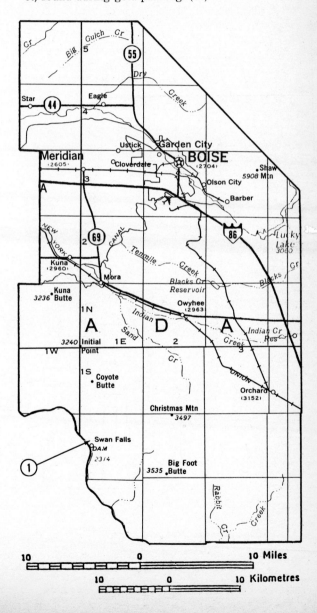

ADAMS COUNTY

1. Meadows area: 2.35 miles east-northeast on Goose Creek;
    at head of Little Goose Creek Canyon.

     Diamond: 3 stones; largest was 0.33 ct; found in 1913.

     Garnet. Ruby. Sapphire. Zircon. (7)

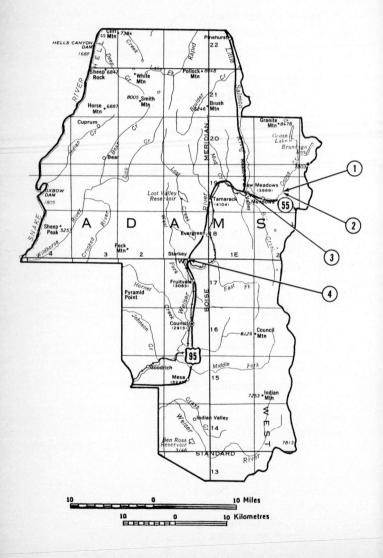

2. Meadows area: 2.3 miles east-northeast on Goose Creek; at the Rock Flat Gold Placer.

   Diamond: 3 gray stones; largest was 1.25 ct. (17)

3. New Meadows area.

   Amethyst. Gold. Topaz. (3-atop bedrock)

4. Starkey area: in location called Seven Devils.

   Brochantite. Epidote: fine crystals to 12″ long. (2)

BANNOCK COUNTY

1. Lava Hot Springs area mines.

   Manganese: with associated minerals. (2)

BEAR LAKE COUNTY

1. Montpelier area: west, in Paris Canyon; at the Humming-bird Mine.

   Quartz, rock: with cuprite inclusions and malachite inclusions; fine; massive. (2)

BENEWAH COUNTY

1. Fernwood area: at Emerald Creek.

   Garnet, almandine. (1-3-7-13-20)

2. Fernwood area: 1.5 miles south-southeast in area where Ruby Gulch meets St. Maries River.

   Garnet, almandine: crystals to 2″ diameter. (1-3-6-7-13-20)

3. Fernwood area: 5 miles southeast; on the East Fork; on the property of Emerald Creek Abrasives (CF).

   Garnet, almandine: marked asterism. (7)

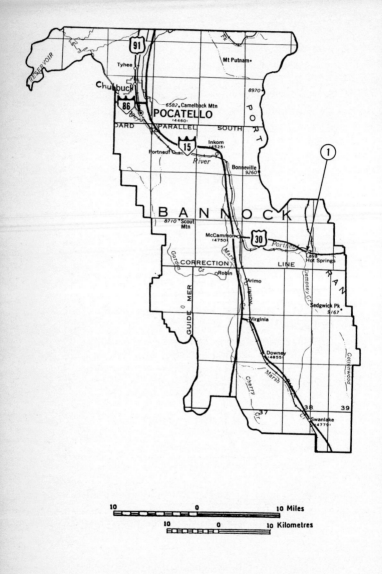

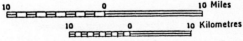

10   0   10 Miles

10   0   10 Kilometres

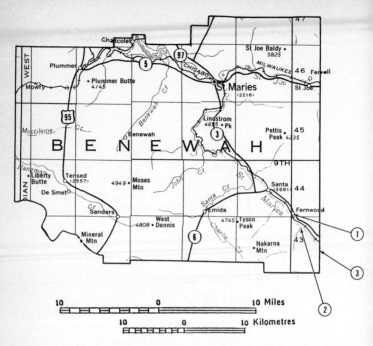

BINGHAM COUNTY

1. Firth area: east; at Sand Creek (aka Willow Creek).
   Petrified wood: fernwood; to 11″ diameter and 2′ length. (1-3-7-13-20)

BLAINE COUNTY

1. Carey area: collect for 8.2 miles north-northeast on the Little Wood River to the Little Wood Reservoir.
   Jasper: fine; red, yellow. (1-7-13-20)

2. Carey area: 1.9 miles to northeast end of Carey Lake; then 7.6 miles upstream in Fish Creek to the Fish Creek Reservoir; collect along all shorelines of reservoir.
   Jasper: fine; red, yellow. Petrified wood. (1-7-13-20)

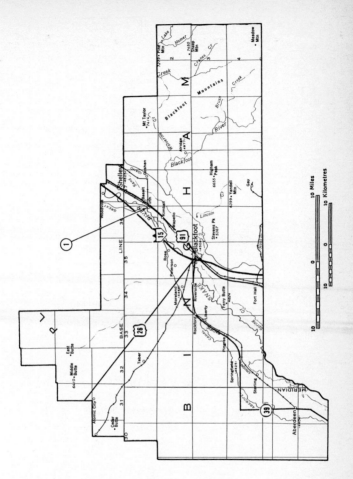

3. Galena area: 0.6 mile south; in headwaters of Big Wood River.

> Chalcedony: geodes to 12″ diameter; fine; some amethyst-lined. (1-14-basalt)

4. Sun Valley area: 12.2 miles east; in headwaters of a branch of the East Fork of Big Wood Creek.

> Chalcedony: geodes to 12″ diameter; fine; some amethyst-lined. (1-14-basalt)

BLAINE COUNTY, IDAHO

DIVIDED INTO TWO
SEPARATE MAPS
-- EACH MAP
IDENTIFIED BY ITS
NUMBER, AS SHOWN
IN DIAGRAM TO LEFT

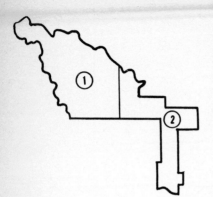

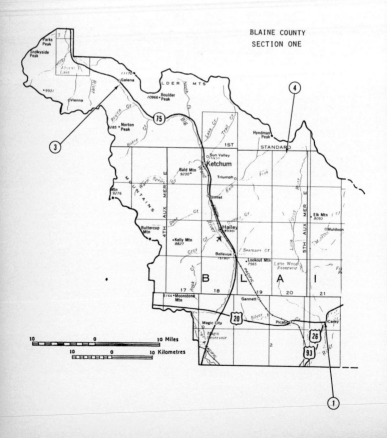

BLAINE COUNTY
SECTION ONE

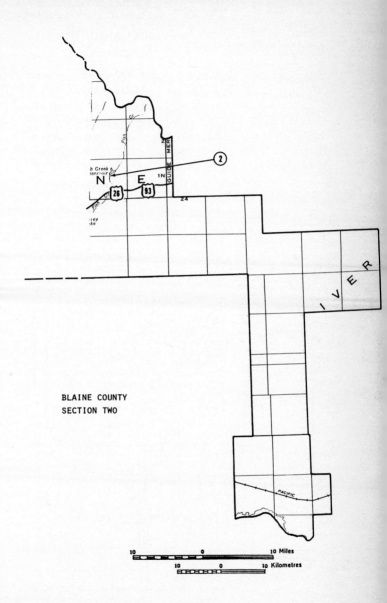

BLAINE COUNTY
SECTION TWO

BOISE COUNTY

1. Centerville area.

     Aquamarine: fine gem quality. (6-16)

2. Idaho City area: on both sides of Mores Creek.

     Opal, precious: in thin films over sandstone. (22)

3. Lowman area: 2 miles downstream on South Fork to mouth of Deadwood Creek; collect upstream (north) in Deadwood Gulch in principal creek and tributary streambeds.

     Garnet, almandine: marked asterism; crystals to 1″ diameter. (3-17)

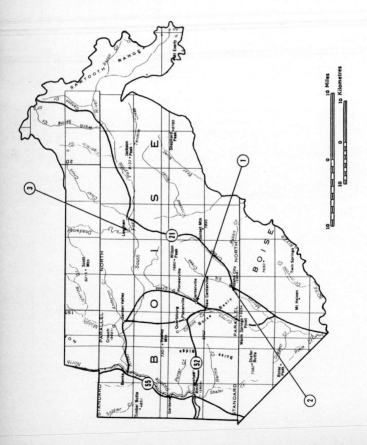

BONNER COUNTY

1. Cabinet area: in mines downstream along Clark Fork.
   Azurite. Chalcocite. Chalcopyrite. Copper. Cuprite.
   Malachite. (2)

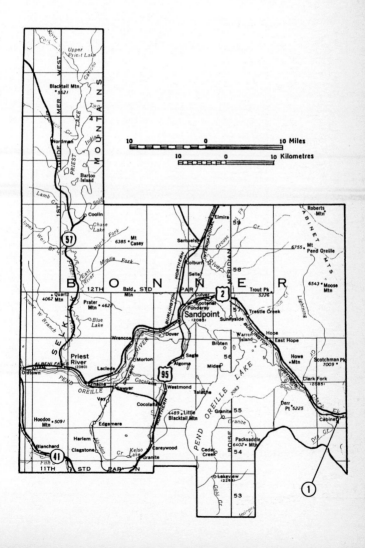

BOUNDARY COUNTY

1. Porthill area: 26 miles southwest on winding road; at series of mines.

> Gold. Lead. Silver. (2)

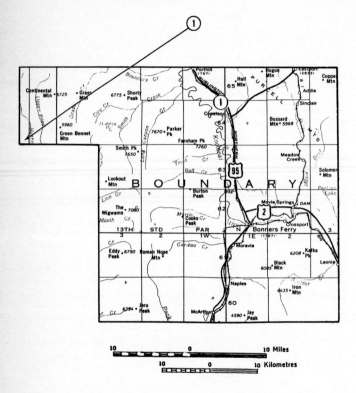

BUTTE COUNTY

1. Arco area: 20 miles west; in series of mines along Lava
Creek.

> Manganese: with associated minerals. Tourmaline. (2)

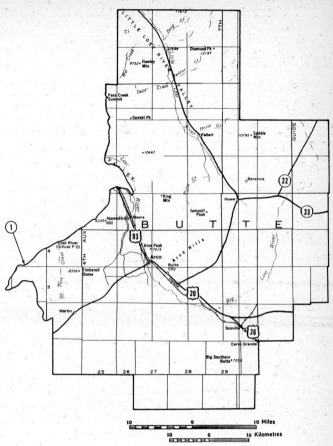

CANYON COUNTY

1. Nampa area: go to the Marsing area and then follow the Homedale Road to fork at the abandoned Y Inn; bear left and go to Graveyard Point; turn left on the canal road and go 1.25 miles (to 0.75 mile before reaching canal bridge).

     Opal, precious: gem quality. (1-13)

2. Nampa area: southwest; at Graveyard Point.

     Agate, plume: with dark ochre marcasite inclusions. (1-3)

3. Nampa area: southwest; in Squaw Creek Canyon; at elevation 3500′; along Squaw Creek just below mouth of Little
   Squaw Creek.

   Chalcedony. Opal, common: milk. Opal, precious: fine
   gem quality. (1-9)

CARIBOU COUNTY

1. Soda Springs area: 3 miles south.
   Gypsum. Sulphur, native: good crystals. (1-13-29)

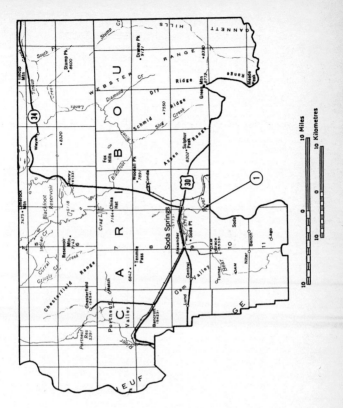

CASSIA COUNTY

1. Moulton area: 5 miles northeast; 3 miles south-southwest of Almo.

    Topaz. (3)

CLEARWATER COUNTY

1. Headquarters area: 1.7 miles upstream (north) in bed of north branch of Reeds Creek.

    Garnet, almandine: asteriate. (3-7-20)

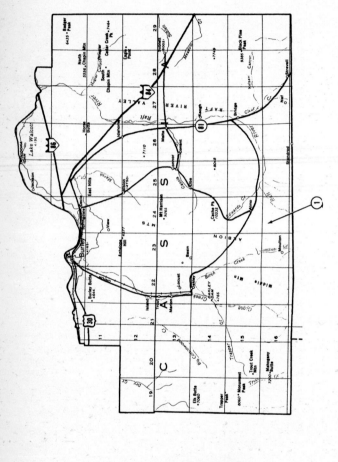

10 Miles
10 Kilometres

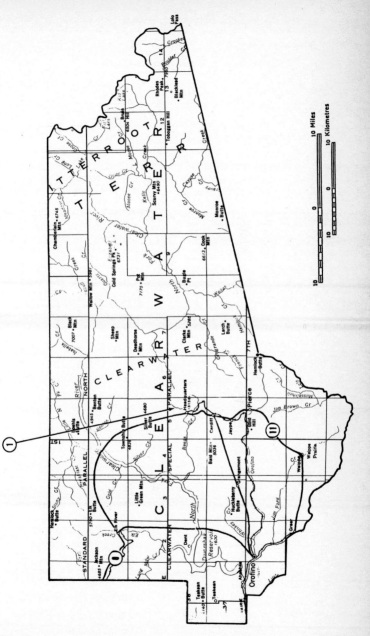

CUSTER COUNTY

1. Bayhorse area.
   Linarite: fine crystals. (1-13)

2. Chilly area: 11 miles southwest; in upper reaches of Big
   Lost Creek valley.
   Chalcedony: geodes to 12″ diameter; fine; some ame-
   thyst-lined. (1-14-basalt)

3. Lost River area: 9.5 miles north-northwest (upstream) in
   Cherry Creek; in headwaters of the creek 0.75 mile south-
   southeast of crest of Shelly Mountain.
   Chalcedony: geodes to 12″ diameter; fine; some amethyst-lined. Jarosite; brown; thick tabular crystals.
   (1-7-13-14-basalt -20)

4. Mackay area: 3 miles west-northwest; 0.5 mile southeast
   of Mackay Reservoir Dam; in the area where North Fork
   empties into Big Lost River and in adjacent hills.
   Chalcedony: geodes to 12″ diameter; fine; some amethyst-lined. (1-14-basalt)

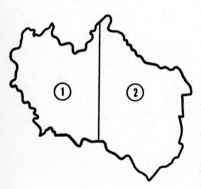

CUSTER COUNTY, IDAHO

DIVIDED INTO TWO
SEPARATE MAPS ON
THE FOLLOWING TWO
PAGES -- EACH MAP
IDENTIFIED BY ITS
NUMBER, AS SHOWN
IN DIAGRAM

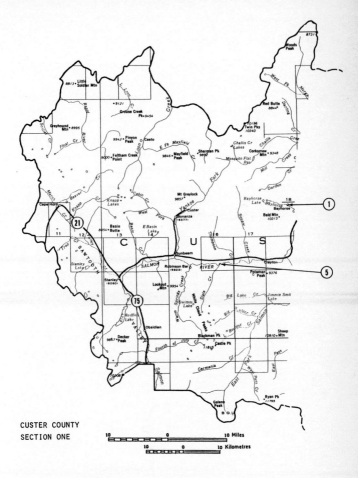

CUSTER COUNTY
SECTION ONE

5. Robinson Bar area: 5.1 miles east on East Fork of Salmon
   River to mouth of Slate Creek; just upstream (south) in
   latter from its mouth.

   Jamesonite. (1-7-13-20)

6. Rothas area: 1.3 miles south; 0.7 mile upstream on Alder
   Creek from where it empties into the North Fork.

   Bronchantite. (1-3-13-20)

CUSTER COUNTY
SECTION TWO

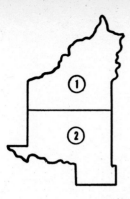

ELMORE COUNTY, IDAHO

DIVIDED INTO TWO
SEPARATE MAPS
-- EACH MAP
IDENTIFIED BY ITS
NUMBER, AS SHOWN
IN DIAGRAM TO LEFT

ELMORE COUNTY

1. Atlanta area mines.
   Gold. (2)

2. Atlanta area: 20 miles west; 6.75 miles north-northwest of
   crest of Sheep Mountain; along North Fork of Boise River
   at a series of old mines.
   Stibnite. (2)

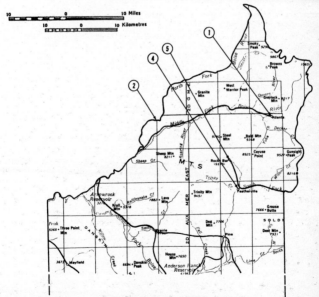

ELMORE COUNTY - SECTION ONE

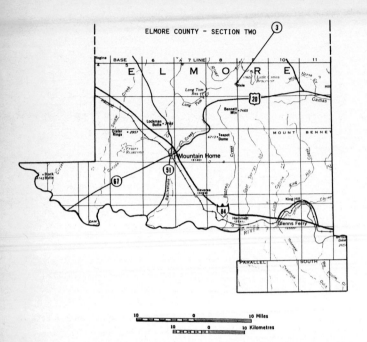

ELMORE COUNTY - SECTION TWO

10   0   10 Miles
10   0   10 Kilometres

3. Dixie area mines.
    Stibnite. (2)

4. Featherville area mines.
    Gold. (2)

5. Rocky Bar area mines.
    Gold. (2)

FREMONT COUNTY

1. St. Anthony area: 18 miles north; 7.9 miles southwest of
    the crest of Bishop Mountain; 3.35 miles northwest of Sand
    Creek Reservoir; at Crystal Butte.
    Andesine: gem quality; crystals to 3.5″ long. (5-14-lava
    flows)

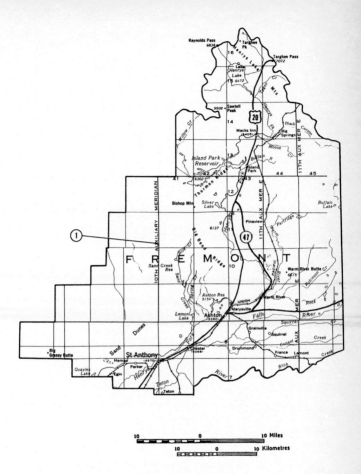

GEM COUNTY

1. Emmett area: 1 mile up Black Canyon on the road to Horseshoe Bend to area of diggings in rock walls.

   Opal, precious: green and red fire; good quality. (1-13-29-rock walls -39)

2. Sweet area: 3.75 miles north-northeast (5 miles east of Squaw Butte).

   Agate. (1-13-39)

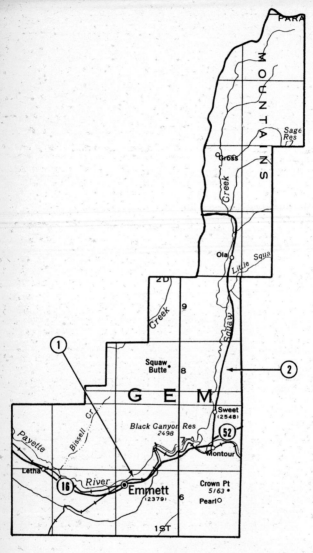

Miles

10   0   10

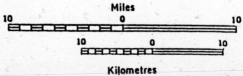

10   0   10

Kilometres

GOODING COUNTY

1. Gooding area: eastward along both sides of Clover Creek
   to the Lincoln County border.

   Opal, wood: some of gem quality. (1-13)

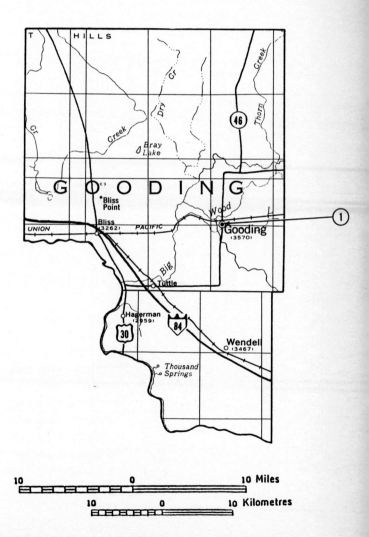

IDAHO COUNTY

1. Grave Butte area: 0.75 mile northeast; in feeder creek of southward-flowing tributary of the Lochsa River.

   Corundum: fine gem quality; tabular crystals; bronze-colored; chatoyant. (3-17)

2. Kooskia area: collect for 3.3 miles south in South Fork, following RR tracks to Stites area.

   Agate. Jasper. (7-18-20)

3. Riggins area and northward toward White Bird; off west side of SR-13/US-95 along 28-mile stretch of Salmon River.

   Agate. Jasper. (1-base of bluffs -7)

4. Slate Creek area.

   Agate. Jasper. (7)

5. Warren area: along tributary of Warren Creek.

   Topaz. (3-17)

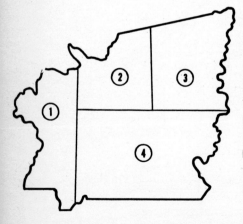

IDAHO COUNTY
IDAHO

DIVIDED INTO FOUR
SEPARATE MAPS ON
THE FOLLOWING FOUR
PAGES -- EACH MAP
IDENTIFIED BY ITS
NUMBER, AS SHOWN
IN DIAGRAM TO LEFT

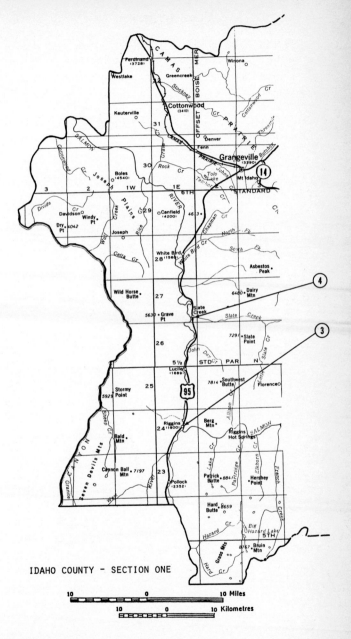

IDAHO COUNTY - SECTION ONE

10   0   10 Miles

10   0   10 Kilometres

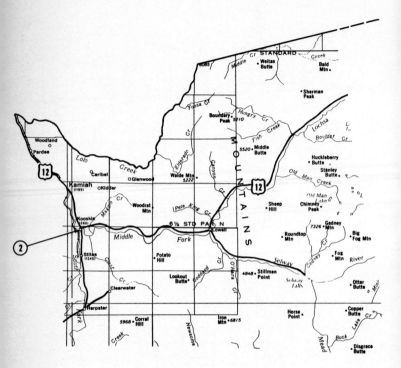

IDAHO COUNTY - SECTION TWO

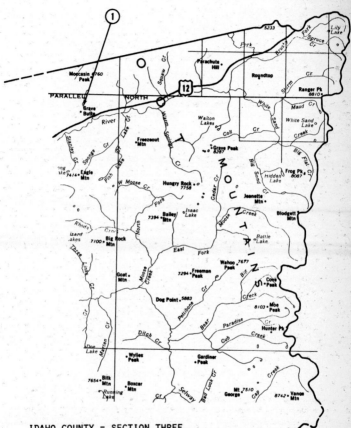

IDAHO COUNTY - SECTION THREE

10 Miles

10 Kilometres

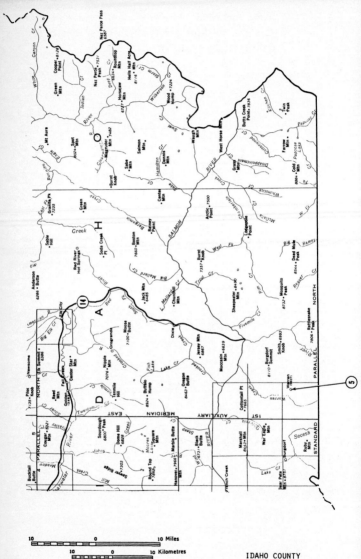

10 |_____| 0 |_____| 10 Miles
10 |_____| 0 |_____| 10 Kilometres

IDAHO COUNTY
SECTION FOUR

KOOTENAI COUNTY

1. Coeur d'Alene area.

   Linarite: well-formed crystals. Pyromorphite: splendid green drusy. (2-38)

2. Setters area.

   Opal, precious: fine gem quality. (5-basalt -29)

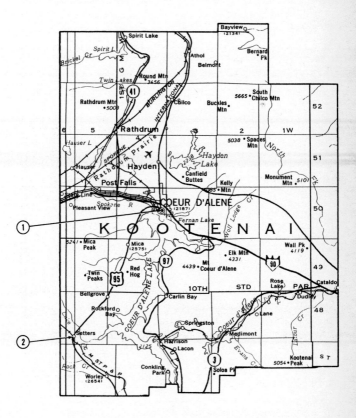

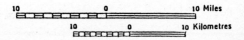

LATAH COUNTY

1. Bovill area: in the St. Joe National Forest's Emerald Creek
   Garnet Area; to reach this go north on SR-3 into Shoshone
   County, past Clarkia to access via Forest Service Road
   447, which goes left from SR-3; follow Forest Service Road
   447 10 miles west to the diggings area. (NOTE: A $4 permit
   must be purchased in order to dig for garnets.)

   Garnet. (1)

2. Moscow area: at ghost town called Gem City on Washing-
   ton-Idaho border.

   Opal, precious: very fine gem quality. (1)

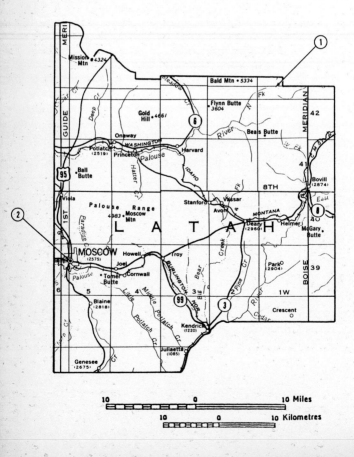

10   0   10 Miles

10   0   10 Kilometres

LEMHI COUNTY

1. Blackbird area: at the Haynes-Stellite Mine.
     Arsenopyrite: fine crystals. Erythrite. Olivenite. Siderite. Vivianite: superb blue-green crystals to 2″ long. (2)

2. Cobalt area: at the Blackbird Mine.
     Vivianite: well-formed crystals; dark green; transparent. (2)

3. Cobalt area: 4.6 miles up Panther Creek; 6 miles below head of creek valley; at elevation 7000′; on the west side of the valley in a large dike paralleling the river.
     Opal, precious: very fine gem quality. (5-porphyry dike)

4. Leesburg area: in Richardson placers.
     Gold. Petrified wood: fine; brown. (7)

5. Lemhi area.
     Huebernite. (2)

6. May area.
     Opal, common: masses; flesh pink, yellow, black. (7)

7. Parker Mountain area.
     Chalcedony. (19)

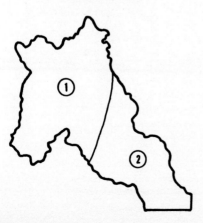

LEMHI COUNTY
IDAHO

DIVIDED INTO TWO
SEPARATE MAPS ON
THE FOLLOWING TWO
PAGES -- EACH MAP
IDENTIFIED BY ITS
NUMBER, AS SHOWN
IN DIAGRAM TO LEFT

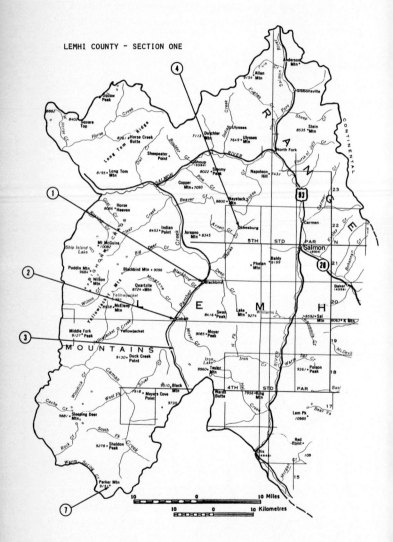

LEMHI COUNTY - SECTION ONE

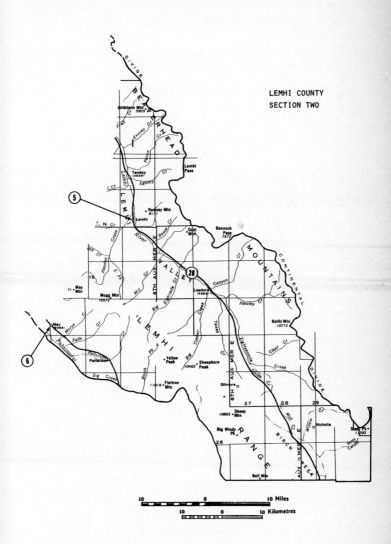

LEMHI COUNTY
SECTION TWO

LEWIS COUNTY

1. Kamiah area: downstream along the west shore of the
   South Fork Clearwater River to the Clearwater County
   border.

   Sillimanite (aka Fibrolite); pale powder blue; as tabular crystals and waterworn pebbles. (7-20)

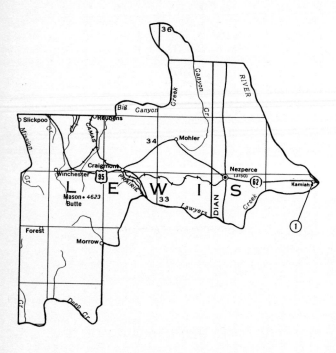

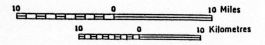

LINCOLN COUNTY

1. Shoshone area: upstream and down along the Wood River.
   Agate. Jasper. (7-18-20)

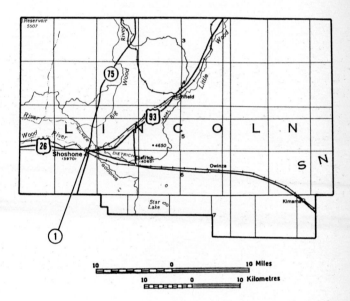

NEZ PERCE COUNTY

1. County-wide.
   Chalcedony: geodes to 12″ diameter. (1-14-basalt)

2. Lewiston area: beginning here and for 100 miles upstream
   along the Clearwater River (into central Clearwater
   County).
   Sillimanite (aka Fibrolite): pale powder blue; as tabular crystals and pebbles. (7-20)

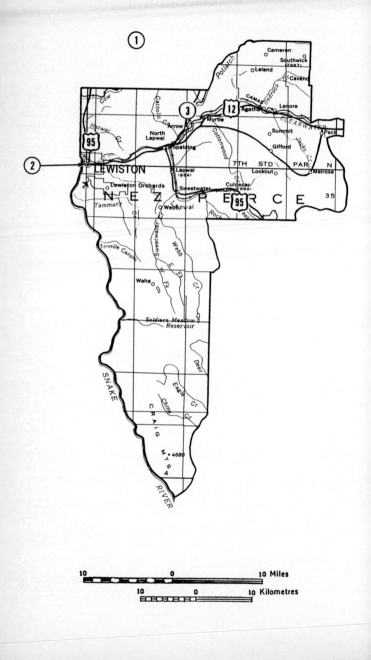

10   0   10 Miles
10   0   10 Kilometres

OWYHEE COUNTY

1. Murphy area: 5.25 miles north; along south shore of Snake River, downstream (northwest) from mouth of Rabbit Creek to mouth of Hardtrigger Creek.

   Chalcedony. Opal, common: milk. Opal, precious: gem quality. (1-9)

2. Oreana area: 5.6 miles east-northeast; 2 miles upstream (south) from Castle Butte; in area of the confluence of Castle Creek and Pickett Creek.

   Opal, precious: fine gem quality. (5-basalt)

3. Reynolds area: in Cliffs district; on the Brace Brothers ranch.

   Opal, precious: gem quality. (5-9-19-basalt)

4. Riddle area: 3.75 miles north to Blue Creek bridge; upstream along Blue Creek.

   Opal, wood: precious; gem quality; usually 4″ to 6″; occasional whole logs; cell structure and annular rings often well preserved. (1-3-5-basalt -6-7-13-20)

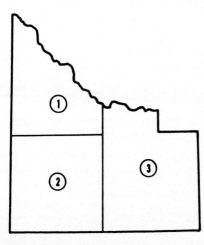

OWYHEE COUNTY
IDAHO

DIVIDED INTO THREE
SEPARATE MAPS ON
THE FOLLOWING THREE
PAGES -- EACH MAP
IDENTIFIED BY ITS
NUMBER, AS SHOWN
IN DIAGRAM TO LEFT

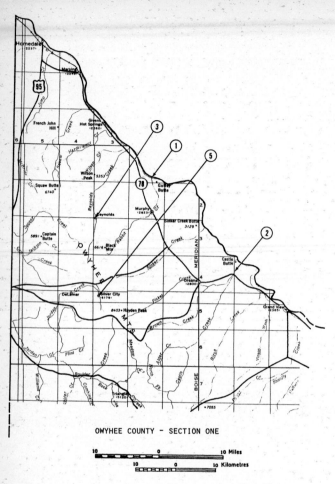

OWYHEE COUNTY - SECTION ONE

5. Silver City area: at the Poorman Mine.

   Adularia: white crystals to 1″ thick. Chloragyrite (aka
   Ceragyrite): well-formed cubes in massive. Pyragy-
   rite. (2)

OWYHEE COUNTY - SECTION TWO

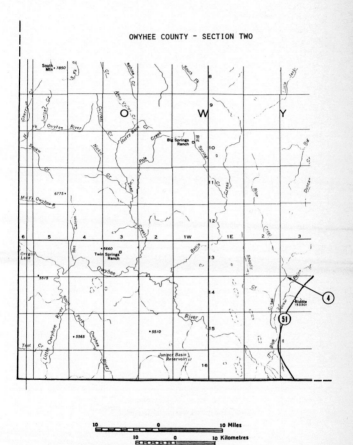

6. Winter Camp area: upstream (southeast) for 30 miles along the East Fork of Bruneau River (aka Clover Creek) to confluence area where Bruneau River is formed by the joining of Big Flat Creek, Three Creek, and Deadwood Creek.

Opal, common. Opal, precious: very fine gem quality. (1-7-13-20)

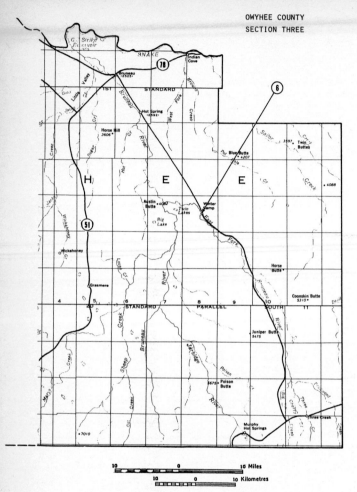

OWYHEE COUNTY
SECTION THREE

10 Miles
10 Kilometres

SHOSHONE COUNTY

1. Avery area: south.
   Staurolite: crystals (often flawed) to 4″ long. (1-3-27)

2. Bathtub Mountain area.
   Staurolite: crystals (usually flawed) to 2″ long. (1-3-27)

SHOSHONE COUNTY
IDAHO

DIVIDED INTO TWO
SEPARATE MAPS
-- EACH MAP
IDENTIFIED BY ITS
NUMBER, AS SHOWN
IN DIAGRAM TO LEFT

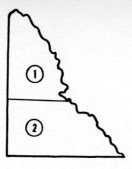

SHOSHONE COUNTY - SECTION ONE

10   0   10 Miles

10   0   10 Kilometres

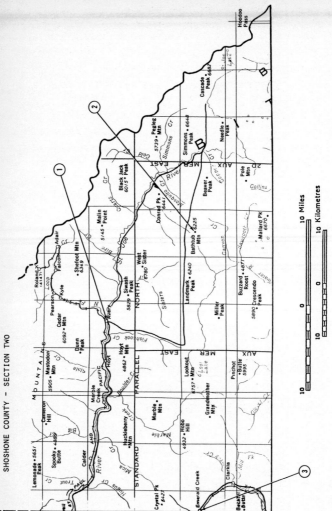

3. Emerald Creek area: along St. Maries River going upstream to RR bridge.

Garnet, almandine: asteriate. (1-3-7-13-18-20)

4. Kellogg area: at the Sunshine Mine.

>   Siderite. Tetrahedrite: lustrous; good crystals to 0.5″
>   thick. (2)

5. Kingston area: at the Hypotheek Mine.

>   Anglesite: well-formed crystals. Galena: massive. (2-5-
>   galena -9)

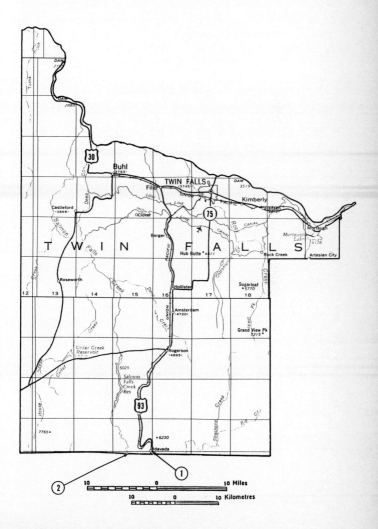

TWIN FALLS COUNTY

1. Idavada area: 4.5 miles north of Jackpot, Nevada; on both sides of US-93, but especially on the west side.
   Chalcedony: geodes to 12″ diameter; some amethyst-lined. (1-7-14-basalt)

2. Idavada area: 4.5 miles north of Jackpot, Nevada; 3.5 miles west of US-93; along banks of Salmon Falls Creek.
   Chalcedony: geodes to 12″ diameter; some amethyst-lined. (1-7-14-basalt)

VALLEY COUNTY

1. McCall area: 2.8 miles west; at Rock Flats.
   Diamond: 19.5 ct; unauthenticated. (17)

2. McCall area: 3 miles west-southwest; in placer operations of Rock Flats.

   Corundum. Gold. Ruby. Sapphire. (3-17)

3. Yellow Pine area: in and around the cinnabar mines.
   Antimony. Opal, common: opaque white with bright red cinnabar inclusions. Scheelite. Stibnite. (1-2)

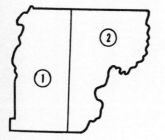

VALLEY COUNTY
IDAHO

DIVIDED INTO TWO
SEPARATE MAPS ON
THE FOLLOWING TWO
PAGES -- EACH MAP
IDENTIFIED BY ITS
NUMBER, AS SHOWN
IN DIAGRAM TO LEFT

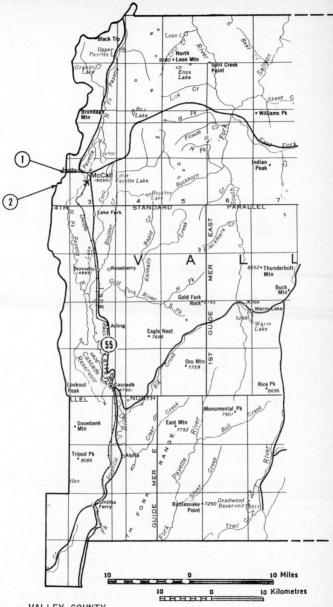

VALLEY COUNTY
SECTION ONE

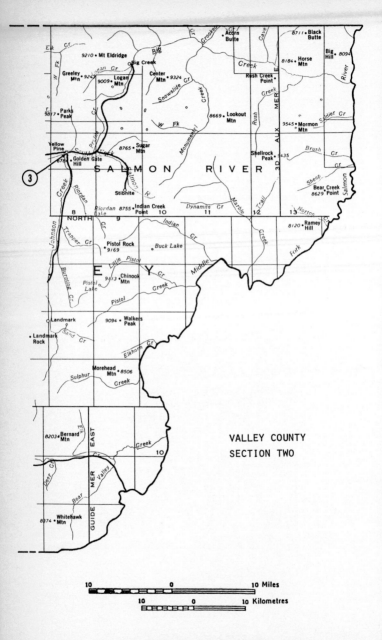

VALLEY COUNTY
SECTION TWO

10   0   10 Miles

10   0   10 Kilometres

WASHINGTON COUNTY

1. Cambridge area: 1.6 miles south to mouth of Little Weiser Creek; then upstream (east) 5 miles above mouth.

   Agate. Chalcedony. (1-7-20)

2. Midvale area: 8.8 miles east; along west base of Hog Creek Butte; in Hog Creek.

   Agate, common. Agate, iris: fine. Chalcedony. (1-7-20)

3. Mineral area: 7 miles southeast of Mineral; in Fourth of July Canyon; 500' above the confluence of Fourth of July Creek with Manns Creek.

   Opal, wood: precious; fine. (6)

4. Weiser area: at Weiser Cove.

   Agate, common. Agate, iris: fine. Chalcedony. Opal, wood: precious; very fine. (1-7)

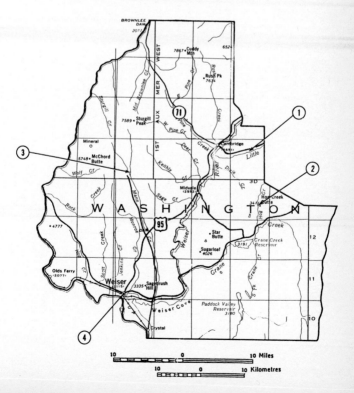

IOWA

Statewide total of 106 locations in the following counties:

Adair (1)	Fayette (2)	Marion (4)
Allamakee (3)	Floyd (2)	Marshall (2)
Appanoose (2)	Franklin (3)	Mitchell (1)
Benton (2)	Fremont (1)	Monroe (6)
Black Hawk (4)	Hamilton (1)	Montgomery (1)
Bremer (1)	Hardin (6)	Muscatine (4)
Buchanan (2)	Henry (8)	Palo Alto (2)
Butler (2)	Jackson (4)	Polk (1)
Cerro Gordo (1)	Jefferson (1)	Scott (1)
Clayton (3)	Johnson (5)	Story (3)
Clinton (1)	Keokuk (2)	Van Buren (2)
Dallas (3)	Lee (4)	Wapello (1)
Des Moines (2)	Linn (2)	Washington (1)
Dickinson (1)	Madison (2)	Webster (1)
Dubuque (3)	Mahaska (2)	

STATEWIDE

Mississippi River (especially fast-flowing areas); also in other rivers, including the Cedar, Wapsipinicon, Maquoketa, Turkey, Iowa, and Skunk, as well as many lakes.

Pearl: abundant (a fine gem pearl of 210 grains was found in the Mississippi River close to shore in the Genoa area). (7)

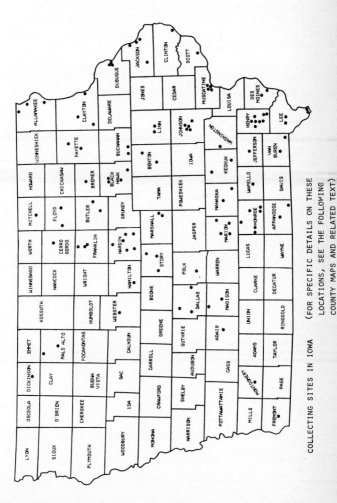

COLLECTING SITES IN IOWA (FOR SPECIFIC DETAILS ON THESE LOCATIONS, SEE THE FOLLOWING COUNTY MAPS AND RELATED TEXT)

ADAIR COUNTY

1. Greenfield area: 0.25 mile east on SR-92 to SR-25; 2.3 miles southeast; collect eastward at base of hills just south of Thompson River.

    Agate. Petrified wood. Quartz, rock. (1-13-25)

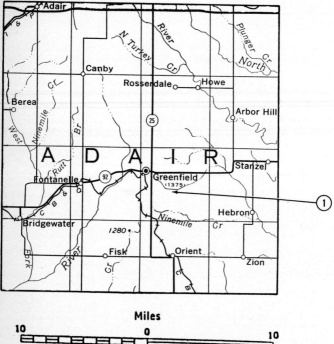

Miles

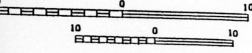

Kilometres

ALLAMAKEE COUNTY

1. Harpers Ferry area: in Mississippi River mud-bed area; probing is especially productive.

    Pearl: gem quality. (7)

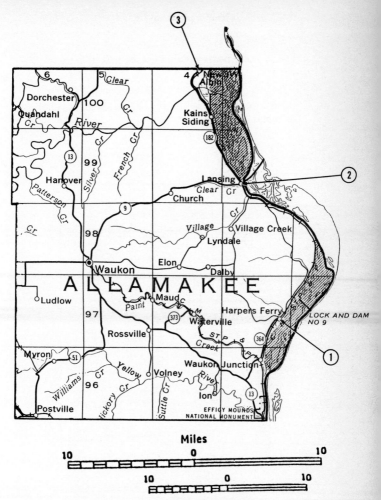

Miles

Kilometres

2. Lansing area: at abandoned mine near town.
   Cerussite. (2)

3. New Albin area mines.
   Galena: good crystals. (2)

APPANOOSE COUNTY

1. Coal City area: in coal mine dumps.
   Pyrite: fine; brassy; cubic crystals to 2″ across. (2)

2. County-wide: in coal mine dumps.
   Pyrite: good crystals; highly brassy. Selenite: sharply
   terminated crystals. (2)

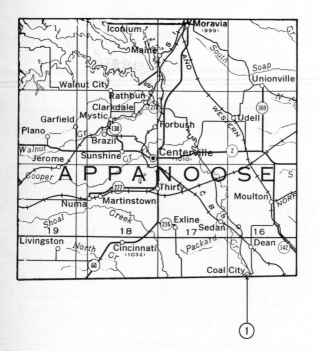

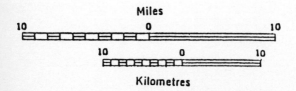

Miles

Kilometres

BENTON COUNTY

1. County-wide.

   Agate, fortification: nodules to 4 lb; some very color-ful; some with cavities lined with sparkling drusy. (1-2-7-13-20-39)

2. Vinton area: north on SR-101; cross the river bridge and then take the first road right; follow it for 4.5 miles along the Cedar River; turn left and go 0.5 mile to gate (CF).

   Agate, fortification. (1-20-39)

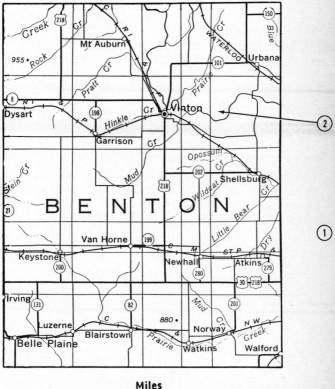

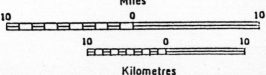

Miles

Kilometres

BLACK HAWK COUNTY

1. County-wide.

   Agate, fortification: nodules to 4 lb; some very colorful; some with cavities lined with sparkling drusy.
   (1-3-4-10-12-13-39)

2. La Porte City gravel pit.

   Agate, banded: nodules to 2″ diameter. (10)

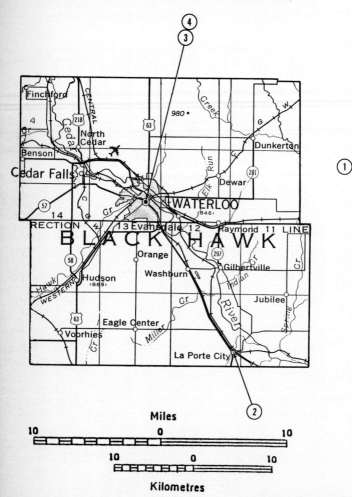

Miles

Kilometres

3. Waterloo area: at the Burton Avenue gravel pit along the Cedar River.

> Agate, banded: nodules to 2″ diameter. Quartz: geodes to 5″ diameter. (10)

4. Waterloo area: at the Mitchell Avenue gravel pit along the Cedar River.

> Agate, banded: nodules to 2″ diameter. Quartz: geodes to 5″ diameter. (10)

BREMER COUNTY

1. County-wide.

> Agate, fortification: nodules to 4 lb; some very colorful; some with cavities lined with sparkling drusy. (4-10-12)

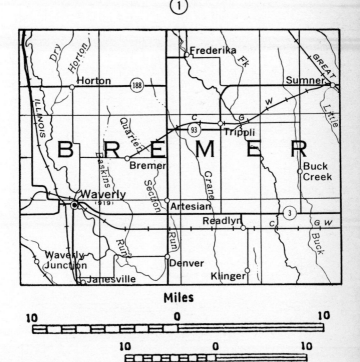

Miles

10   0   10

10   0   10

Kilometres

BUCHANAN COUNTY

1. Brandon area: 0.6 mile east on SR-283 to the Lime Creek bridge; collect upstream and down in Lime Creek and its tributaries.

   Chalcedony: geodes to 8″ diameter. Chalcedony: nodules to 8″ diameter. Fossils, marine invertebrates. (1-7-13-20)

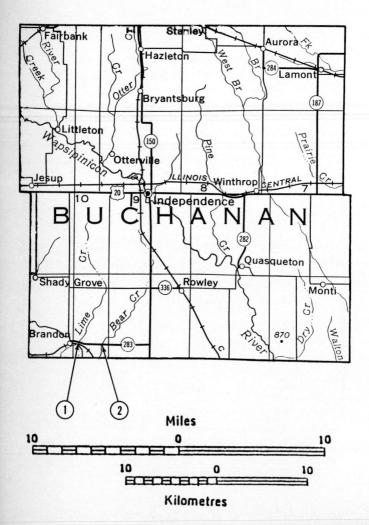

Miles

Kilometres

2. Brandon area: 2.1 miles east on SR-283 to bridge over Bear
   Creek; collect upstream and down in Bear Creek and its
   tributaries.

   Chalcedony: geodes to 8″ diameter. Chalcedony:
   nodules to 8″ diameter. Fossils, marine invertebrates.
   (1-7-13-20)

BUTLER COUNTY

1. New Albion area: 0.8 mile north to Cedar River; then collect for 8.2 miles upstream (west) to mouth of West Fork
   (the third tributary upstream).

   Agate, banded. Quartz: geodes to 6″ diameter; some
   with drusy-lined cavities. (1-7-13-20)

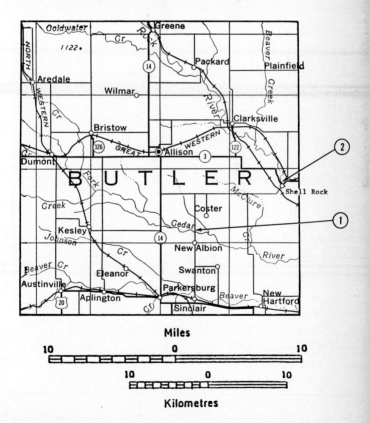

Miles

10   0   10

10   0   10

Kilometres

2. Shell Rock area: north; just beyond town limits, at gravel pit.

> Agate, banded. (10)

CERRO GORDO COUNTY

1. Mason City area quarries.

> Fossils: marine; primarily corals, crinoid stems, gastropods, pelecypods, and occasional trilobites. (12)

CLAYTON COUNTY

1. Elkader area gravel pit along Turkey River.

    Agate, banded. Jasper: black, red. Quartz: geodes to 8″ diameter. (10)

2. Guttenberg area.

    Agate, banded: gray. Calcite ("cave onyx" var.). Jasper: black. (1-13-14-limestone bluffs)

3. Strawberry Point area.

    Agate, fortification (Lake Superior var.): large (including an 8-lb nodule found in 1979). (1-3-10-13)

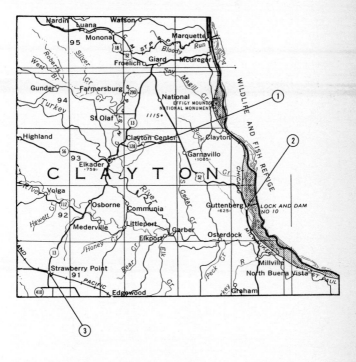

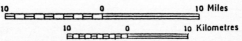

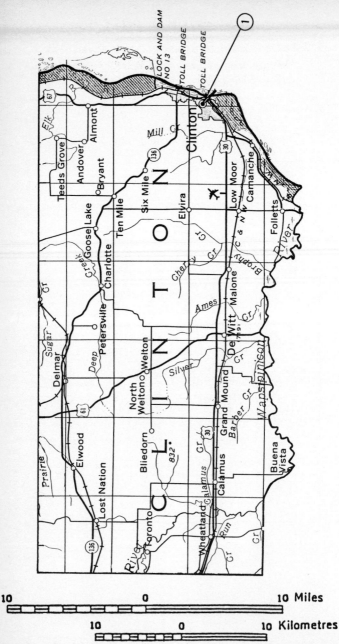

LOCK AND DAM NO 13
TOLL BRIDGE
TOLL BRIDGE
①

Elk
Teeds Grove
Andover
Almont
Bryant
67

Mill Cr.
Clinton
136
30

C
L
I
N
T
O
N

Ten Mile
Six Mile
Elvira
Low Moor
Camanche
Goose Lake
Charlotte
Cherry Cr.
Cr.
C & N W
Broph

Follets

Creek

Cr.

River

Petersville
Ames
De Witt
Malone
Cr.
791

Sugar
Cr.
Delmar
Deep
North Welton
Welton
Silver
Grand Mound
Cr.
Barber Cr.
Wapsipinicon
61

Bliedorn
832
Calamus Cr.
30
Buena Vista

Prairie
Elwood
Lost Nation
Toronto
Wheatland
Calamus
Run
Cr.
Cr.

River
136

10	0	10 Miles

10	0	10 Kilometres

CLINTON COUNTY

1. Clinton area: at River Drive and Fourth Street; in the Block Company gravel operation.

   Agate, banded: nodules to 2″ diameter. Carnelian. Petrified wood: waterworn. (10)

DALLAS COUNTY

1. Dallas Center area: 7 miles east; then 1.25 miles north; in gravel operation.

   Agate, banded. Quartz: geodes to 7″ diameter. (10)

2. Dawson area: 1.9 miles south on country road to Elm Branch bridge; collect downstream 6 miles to mouth of branch at Raccoon River.

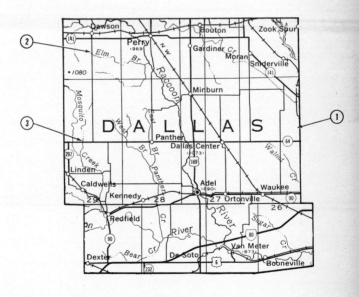

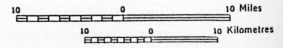

Agate, banded. Quartz: geodes to 9″ diameter. (1-7-13-20)

3. Linden area: 2.8 miles north on SR-292 to right turn on SR-64; pass over first bridge (Mosquito Creek) and stop at second bridge, which is intermittent tributary of Mosquito Creek; collect upstream (north) for 3 miles.

Quartz: geodes to 8″ diameter. (1-7-13-20)

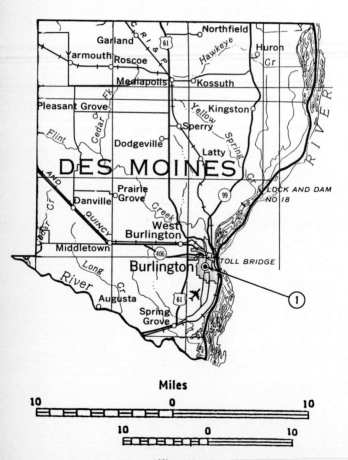

Miles

Kilometres

DES MOINES COUNTY

1. Burlington area gravel-dredging operations and gravel pits.

    Agate: various forms, including banded, bull's-eye (orbicular), clear, colored (opaque and transparent), fortification (Lake Superior), moss, and oolitic; nodules to 3″ diameter. Carnelian. Fossils: marine; chiefly corals; some silicified. Jasper. Moonstone. Petrified wood: waterworn. (10)

2. County-wide.

    Chalcedony: geodes to 9″ diameter. Quartz: geodes to 9″ diameter. (1-3-4-6-7-10-12-13-18-22-28-29)

DICKINSON COUNTY

1. Okoboji area: along the shoreline of West Okoboji Lake.

    Sapphire: blue; fine gem quality; many flawless; one cut to fine gemstone of 1.375 ct. (20)

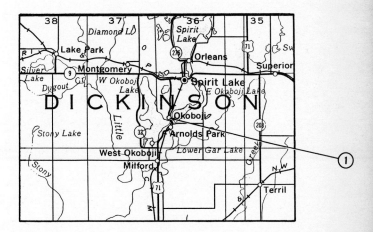

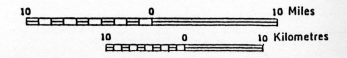

DUBUQUE COUNTY

1. Dubuque area gravel-dredging operations.

   Agate: various forms, including fortification, moss, and oolitic. Fossils: marine; chiefly corals; some silicified. Jasper: black, red, yellow, occasional green. Moonstone. Petrified wood: waterworn. (10)

2. Dubuque area: just north of US-20 on East Third Street; at the Molo Sand and Gravel Company.

   Agate, banded: nodules to 2″ diameter. Carnelian. Jasper: scarce. (10)

3. Dubuque area: north of city; just west of YMCA camp in Eagle Park.

   Calcite ("satin spar" var.): chatoyant. (1-8-14)

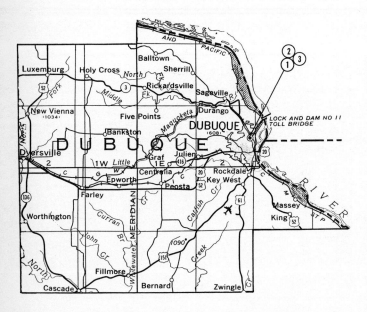

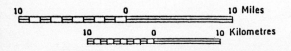

FAYETTE COUNTY

1. County-wide: especially in streams.

   Agate, fortification; nodules to 4 lb; some very colorful; some with cavities lined with sparkling drusy.
   (1-3-4-7-13-20-39)

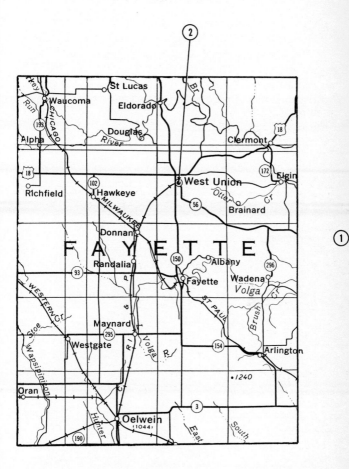

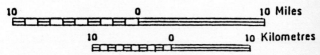

| 10 | 0 | 10 Miles |
| 10 | 0 | 10 Kilometres |

2. West Union area gravel pit.

> Agate, banded; nodules to 2″ diameter. Agate, fortification; nodules to 4 lb. (10)

FLOYD COUNTY

1. Charles City area quarries.

> Fossils: marine; primarily corals, crinoid stems, gastropods, pelecypods, occasional trilobites. (12)

2. Rockford area: 0.5 mile west on SR-147; in series of holes and dumps of the Rockford Brick and Tile Company; on the Shellrock River.

> Fossils, marine: especially brachiopods, bryozoans, corals. (10-12-21)

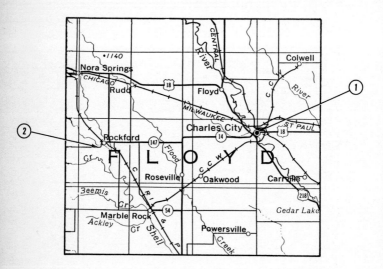

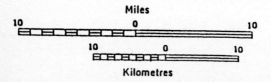

FRANKLIN COUNTY

1. Chapin area: in local quarry.

    Chalcedony: geodes to 4″ diameter; often having cavities lined with sparkling drusy. Quartz: geodes to 4″ diameter; often having cavities lined with small, well-terminated crystals. (5-limestone -12)

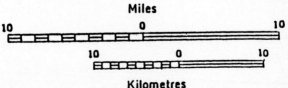

2. Chapin area: northward on both sides of US-65 for 3.8 miles, to the Sheffield area.

   Quartz: geodes to 6″ diameter. (1-7-13)

3. Sheffield area: in quarry and fields behind quarry.

   Quartz: geodes to 5″ diameter. (1-12-13)

FREMONT COUNTY

1. County-wide in stream valleys.

   Fossils, marine: good corals; stromatoporoids. (1-7-13-22-29-limestone)

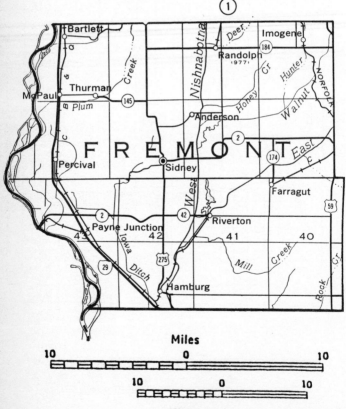

Miles

Kilometres

HAMILTON COUNTY

1. Stanhope area: north on SR-17/60 to second crossroad;
   turn left and drive west to end of maintained road; north
   from here on well-beaten car trail to bridge; on left in
   exposed veins.

   Calcite: unusual cone-in-cone formations; to 12″ long;
   fragile; black. (19-22-25-28)

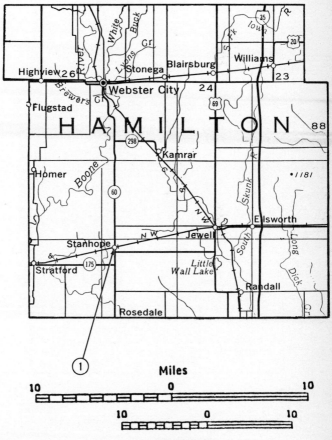

HARDIN COUNTY

1. Buckeye area: 2.4 miles south on country road running along west side of RR tracks to bridge over Tipton Creek; collect downstream on both sides of creek to where it empties into South Fork of Iowa River east-southeast of Point Pleasant.

    Quartz: geodes to 6″ diameter. (1-7-13-20)

2. Gifford area: 0.35 mile north on SR-215 to bridge over South Fork of Iowa River; collect upstream on both sides all the way to vicinity of Wilke.

    Quartz: geodes to 6″ diameter. (1-7-13-20)

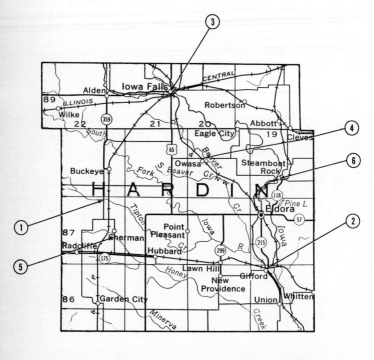

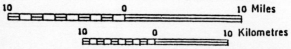

3. Iowa Falls area: upstream along both shores of Iowa River to Franklin County border.

   Quartz: geodes to 6″ diameter. (1-13-20)

4. Owasa area: 0.25 mile east to Beaver Creek; collect downstream along both sides to where it empties into South Fork of Iowa River, 2 miles northwest of Gifford.

   Quartz: geodes to 6″ diameter. (1-7-13-20)

5. Sherman area: 0.25 mile south on country road to Honey Creek; collect downstream on both sides to Story County border.

   Quartz: geodes to 6″ diameter. (1-7-13-20)

6. Steamboat Rock area: downstream along both shores of Iowa River to Marshall County border.

   Quartz: geodes to 5″ diameter. (1-7-13-20)

HENRY COUNTY

1. County-wide.

   Chalcedony: geodes to 9″ diameter. Quartz: geodes to 9″ diameter. (1-3-4-6-7-10-12-13-18-20-22-28-29)

2. Lowell area: at Geode State Park near Skunk River.

   Quartz: geodes to 12″ diameter; mainly milky; some cavities amethystine-lined. (1-3-13-14)

3. Lowell area: east to bridge over Mud Creek; upstream in creek.

   Agate. Quartz: geodes to 10″ diameter. (1-7-13-20)

4. Mount Pleasant area: 2.4 miles west on US-34; in gravel pit adjacent to Big Creek.

   Agate, banded. (10)

5. Mount Pleasant area: 3.9 miles west on US-34; in gravel pit adjacent to Skunk River.

   Agate, banded. (10)

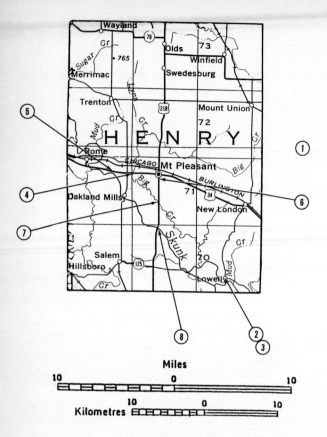

6. Mount Pleasant area: southern edge of town.

   Chert, banded: fine; white, ivory, red, brown, gray,
   dark grayish blue. (5-limestone -14-29)

7. Mount Pleasant area: 2.3 miles south on US-218 to bridge
   over Big Creek; collect upstream and down on both sides
   of Big Creek.

   Agate, banded. (7-13-20)

8. Mount Pleasant area: 4.5 miles south on US-218 to bridge
   over Skunk River; collect upstream on both sides of Skunk
   River.

   Agate, banded. (7-18-20)

JACKSON COUNTY

1. Bellevue area: along Mississippi River at gravel-dredging operations.

   Agate: various forms, including fortification, moss, and oolitic. Fossils: marine; chiefly corals; some silicified. Jasper: red, yellow. Moonstone. Petrified wood: waterworn. (10)

2. Bellevue area: 0.25 mile north on US-52 to gravel pit on right; gravel in large washed piles; larger nodules occasionally in oversized gravel discard pile near loading dock.

   Agate, banded. Agate, fortification. Carnelian. Petrified wood: waterworn. (10)

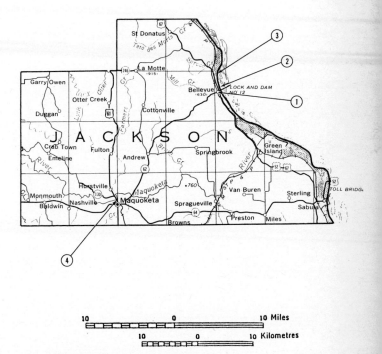

10   0   10 Miles

10   0   10 Kilometres

3. Bellevue area: 1.4 miles north of north town limit sign on US-52; at the Bellevue Sand & Gravel Company on right.

> Agate, banded. Agate, fortification. Fossils: marine; chiefly coral. Jasper: black, red, yellow, green. Petrified wood: waterworn. (10)

4. Maquoketa area: just north on US-61 at quarry.

> Fossils: marine; especially good brachiopods. (12)

JEFFERSON COUNTY

1. County-wide.

> Chalcedony: geodes to 7″ diameter. Quartz: geodes to 7″ diameter. (1-3-4-6-7-10-12-13-18-20-22-28-29)

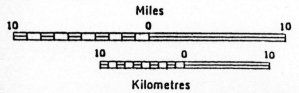

JOHNSON COUNTY

1. Coralville area gravel pits.

  Fossils: marine; chiefly corals; many silicified. (10)

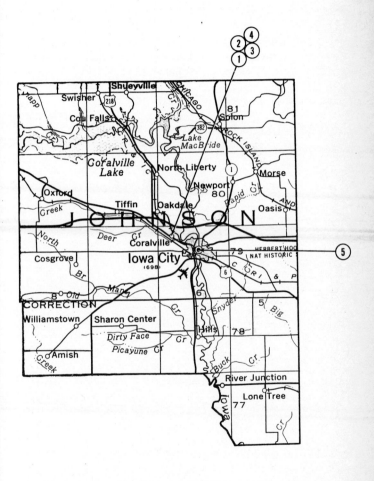

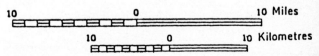

2. Coralville area quarries.

> Fossils: marine; chiefly corals; silicified. (12)

3. Coralville area streambeds.

> Agate, banded. Fossils: marine; chiefly corals; occasionally silicified. (7-20)

4. Coralville area bluffs.

> Calcite. Fossils: marine; chiefly corals; some silicified. Quartz: geodes to 8″ diameter. (1-4-6-13-14)

5. Iowa City area stream bluffs.

> Calcite. Fossils: marine; chiefly corals; some silicified. Quartz: geodes to 9″ diameter. (3-7-14-20)

KEOKUK COUNTY

1. Keswick area quarry.

> Agate, banded: nodules to 1.5″ diameter. (12)

2. Ollie area: 1 mile west; then 0.3 mile north to quarry.

> Agate, banded. (12)

LEE COUNTY

1. County-wide.

> Chalcedony: geodes to 9″ diameter. Quartz: geodes to 9″ diameter. (1-3-4-6-7-10-12-18-20-22-28-29)

2. Donnellson area: just south of town in gravel pit.

> Quartz; geodes to 8″ diameter. (10)

3. Keokuk area: large region (100 miles north to south and 60 miles east to west); in road cuts and fields and especially in streambeds throughout this region of southeastern Iowa, western Illinois, and northeastern Missouri.

> Calcite: geodes to 9″ diameter. Chalcedony: geodes to 9″ diameter. Quartz: geodes to 9″ diameter. (1-4-6-7-13-20)

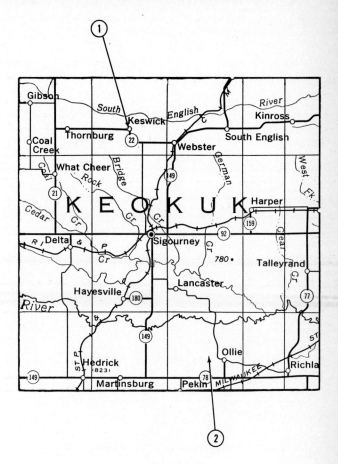

K E O K U K

Gibson
South
Keswick
English
River
Kinross
Thornburg
② 22
Webster
South English
Coal
Creek
What Cheer
Bridge
German
West Fk
② 21
Rock
Cr
Harper
Cedar
Gr
Cr
② 159
R / Delta
&
P
Cr
Sigourney
Cr
② 92
Cr
Hayesville
② 180
Lancaster
780 •
Talleyrand
② 77
River
&
② 149
Ollie
ST
Hedrick
(823)
② 149
Martinsburg
Pekin
② 78
MILWAUKEE
Richla
ST

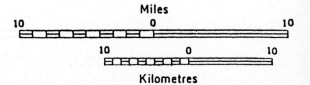

Miles
10   0   10

10   0   10
Kilometres

4. Keokuk area: south on Fifth Street to the Union Carbide plant; collect just west of the plant.

Quartz: geodes to 8″ diameter. (28-upper)

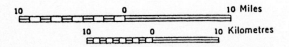

LINN COUNTY

1. Cedar Rapids area gravel pits.
   > Agate, banded. Agate, fortification. Jasper: black, yellow. (10)

2. County-wide.
   > Calcite (fossil coral). (1-3-7-13-20)

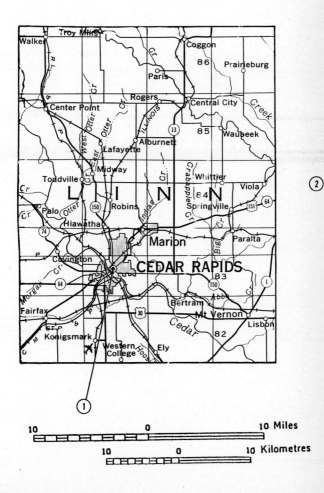

MADISON COUNTY

1. Earlham area quarries.
   Fossils: marine; primarily invertebrates. (12)

2. Winterset area quarries.
   Fossils: marine; primarily invertebrates. (12)

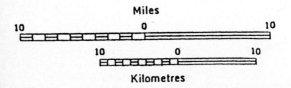

MAHASKA COUNTY

1. Cedar area: southeast along both sides of Cedar Creek to its mouth at Skunk River.

    Fossils, marine invertebrates: especially corals; some calcified; some silicified. (1-7-13-20-29-limestone)

2. Oskaloosa area coal mine dumps.

    Calcite: long, white, sharply terminated crystals in clusters. Dolomite: fine crystals; iridescent bronzy greenish black. Marcasite. Pyrite. (2)

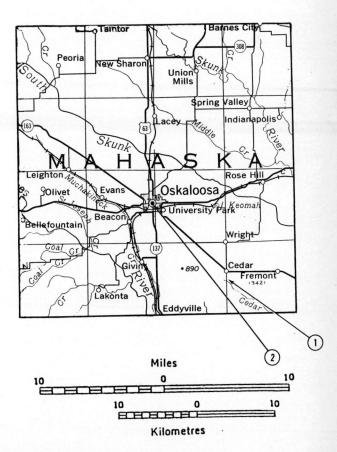

Miles

Kilometres

MARION COUNTY

1. County-wide in coal mine dumps.

    Barite: fine crystals. Calcite: white; sharply ter-
    minated crystals in clusters. Dolomite: bronze crys-
    tals. Marcasite. Pyrite. Siderite: crystals. (2-41)

2. Knoxville area strip mines.

    Calcite: spectacular tiny specimens; brown; iridescent;
    encrusted with delicate acicular snow-white crystal
    blossoms. (2)

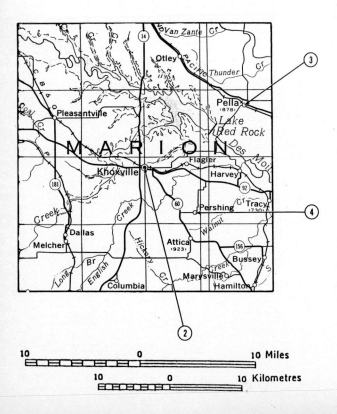

10   0   10 Miles

10   0   10 Kilometres

3. Pella area quarry.

   Fossils: marine; chiefly blastoids, brachiopods, horn corals, small black trilobites. (12)

4. Pershing area: at the Pershing Mine dump.

   Barite: crystals. Calcite: white; sharply terminated crystals in clusters. Dolomite: crystals. Fossils: marine; chiefly brachiopods and corals. Marcasite: good crystals. Pyrite: high-quality brassy crystals. Siderite: crystals. (2-14)

MARSHALL COUNTY

1. Quarry area: 1.8 miles northeast across the Iowa River bridge to the Concrete Materials Company quarry.

   Fossils: marine; very fine, including whole starfish and other echinoids. (12)

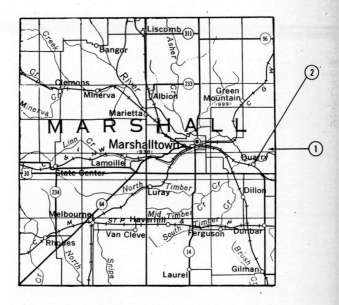

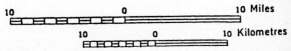

2. Quarry area: at old quarry along Iowa River.
   Fossils: marine; some very fine. (12)

MITCHELL COUNTY

1. Osage area.
   Agate, fortification (Lake Superior var.): fine; some quite large (the third largest on record, weighing 32 lb, was found here). (1-3-7-10-13-18)

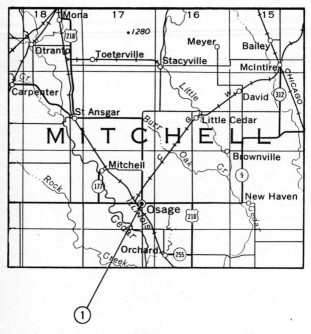

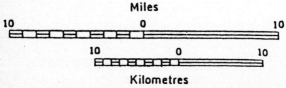

Miles

10   0   10

10   0   10

Kilometres

MONROE COUNTY

1. Albia area coal mine dumps.
   Pyrite: good crystals. Selenite: sharply terminated crystals. (2)

2. Brompton area coal mine dumps.
   Pyrite: good crystals. Selenite: sharply terminated crystals. (2)

3. Foster area coal mine dumps.
   Pyrite: good crystals. Selenite: sharply terminated crystals. (2)

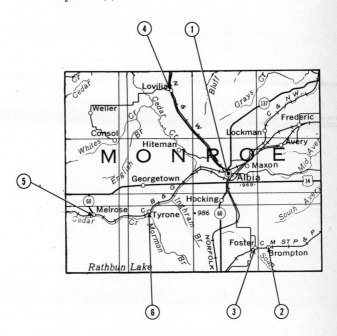

Miles

Kilometres

4. Lovilia area coal mine dumps.

> Calcite: good crystals with goethite inclusions. Goethite: fine; hairlike crystals; most often as calcite inclusions. (2)

5. Melrose area coal mine dumps.

> Calcite: white; good crystals; sharply terminated; usually in clusters; often with goethite inclusions. Goethite: fine; hairlike crystals; usually as calcite inclusions. Marcasite. Pyrite: good crystals. Selenite: sharply terminated crystals. (2)

6. Tyrone area coal mine dumps.

> Calcite: white; good crystals. Marcasite. Pyrite: brassy cubic crystals; usually striated. Selenite: sharply terminated crystals. (2)

MONTGOMERY COUNTY

1. Red Oak area: 1 mile west on US-34 to side road right; go north a short distance to limestone quarry on left.

> Chert (Rice Agate var.): cutting grade; from 2″ to 2′ thick. (12-19)

MUSCATINE COUNTY

1. Muscatine area: at the Acme Company gravel-dredging operation.

> Agate: various forms, including banded, fortification, moss, and oolitic. Fossils: marine; chiefly corals; many silicified. Jasper: black, red, yellow, some green. Moonstone. Petrified wood: waterworn. (10)

2. Muscatine area: at the Hahn Brothers Gravel Company.

> Agate: various forms, including banded, fortification, moss, and oolitic. Fossils: marine; chiefly corals; some silicified. Jasper: black, red, some yellow. Moonstone. Petrified wood: waterworn. (10)

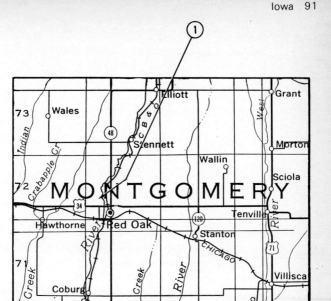

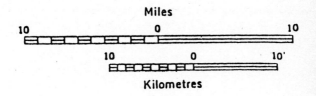

3. Muscatine area: at the Northern Gravel Company.

   Agate: various forms, including banded, fortification, moss, and oolitic. Fossils: marine; chiefly corals; some silicified. Jasper: black, red, yellow, some green. Moonstone. Petrified wood: waterworn. (10)

4. Muscatine area: south.

   Agate, fortification: nodules to 8″ diameter; good quality and color. (1-3-26)

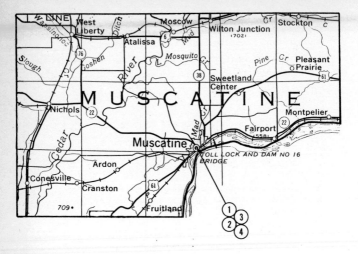

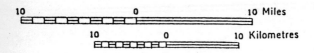

PALO ALTO COUNTY

1. Emmetsburg area.
      Petrified wood. (3)

2. Graettinger area.
      Petrified wood. (3)

POLK COUNTY

1. Des Moines area strip coal mines.
      Petrified wood. Pyrite. (2)

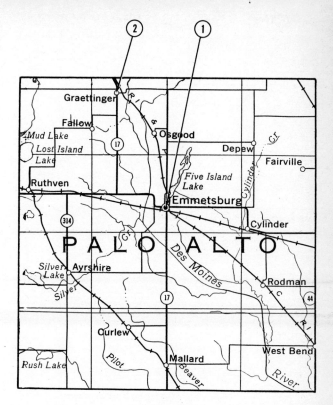

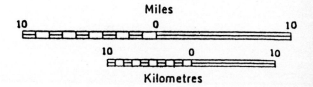

SCOTT COUNTY

1. Davenport area gravel pits.

    Agate, banded: nodules to 2″ diameter. Carnelian. Petrified wood. (10)

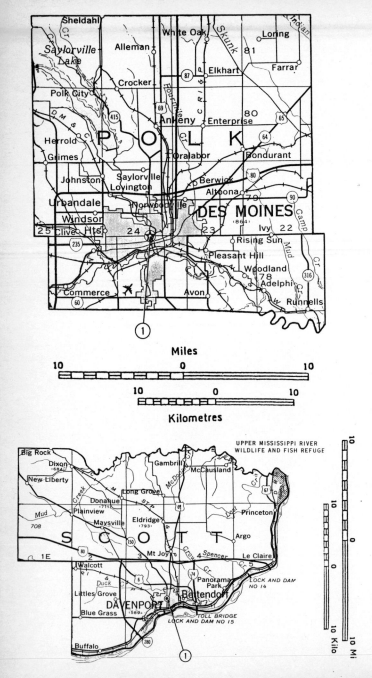

Miles

Kilometres

UPPER MISSISSIPPI RIVER
WILDLIFE AND FISH REFUGE

STORY COUNTY

1. Ames area quarries.
   Fossils: marine; primarily invertebrates. (12)

2. McCallsburg area quarries.
   Quartz: geodes to 4″ diameter. (5-limestone -12)

3. Nevada area: 3.5 miles east on US-30 to bridge over East
   Indian Creek; collect upstream on both sides.
   Agate, banded. Agate, fortification. Fossils: marine.
   Jasper: red. (1-7-13-18-20)

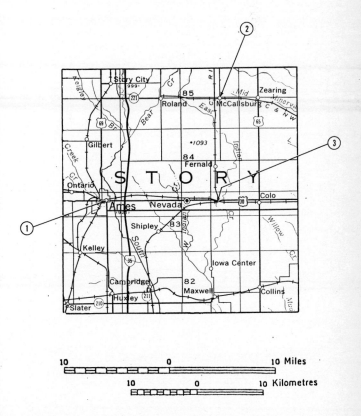

VAN BUREN COUNTY

1. County-wide.

    Chalcedony: geodes to 8″ diameter. Quartz: geodes to
    8″ diameter. (1-3-4-6-7-10-12-18-20-22-28-29)

2. Farmington area.

    Quartz: geodes to 7″ diameter. (1-13-14-20)

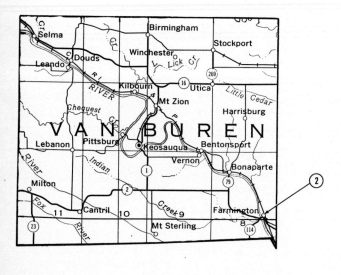

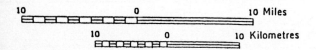

WAPELLO COUNTY

1. Ottumwa area: southeast in cut banks along both sides of
   the Des Moines River to the Davis County border.

    Fossils, marine: especially corals; some calcified; some
    silicified. (1-7-13-20-29-limestone)

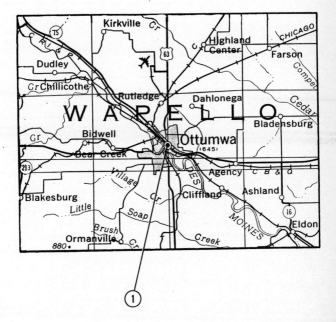

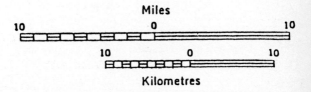

Miles

10   0   10

10   0   10

Kilometres

WASHINGTON COUNTY

1. Keota area: at the Kaser Construction Company quarry.
   Agate, banded. (12)

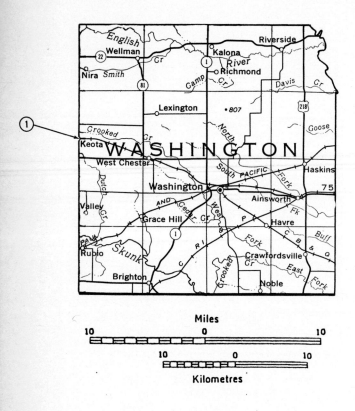

Miles

10   0   10

10   0   10

Kilometres

WEBSTER COUNTY

1. Fort Dodge area quarries.
   Gypsum: good crystals. (9-12)

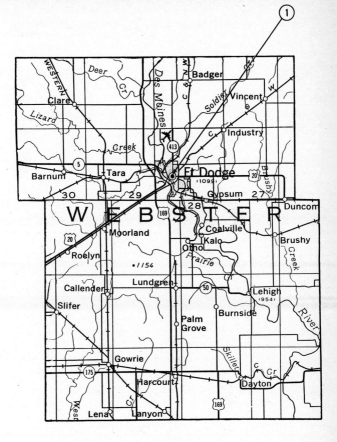

WESTERN
Deer
Cr.
Badger
Clare
Lizard
Soldier
Vincent
Creek
Des Moines
Industry
413
5
Barnum
Tara
Ft Dodge
(1099)
20
30
29
Gypsum 27
Duncom
WEBSTER
28
169
Moorland
Coalville
20
Kalo
Brushy
Othol
Roelyn
Prairie
Creek
•1154
Lundgren
50
Callender
Lehigh
(954)
Slifer
Burnside
River
Palm
Grove
Skillet
Gowrie
Cr
C
175
Dayton
Harcourt
169
West
Cr
Lena
Lanyon

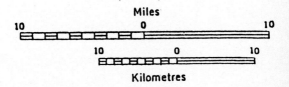

Miles
10   0   10

10   0   10
Kilometres

KANSAS

Statewide total of 81 locations in the following counties:

Anderson (1)	Franklin (2)	Osborne (2)
Barber (5)	Geary (2)	Ottawa (2)
Bourbon (1)	Gove (2)	Reno (1)
Brown (1)	Graham (1)	Rice (2)
Butler (2)	Harper (2)	Riley (1)
Chase (2)	Jefferson (1)	Russell (1)
Cherokee (2)	Kingman (1)	Saline (1)
Clark (3)	Kiowa (1)	Shawnee (1)
Cloud (2)	Logan (2)	Sheridan (2)
Comanche (2)	Marion (1)	Trego (4)
Cowley (2)	Marshall (1)	Wallace (7)
Dickinson (2)	McPherson (2)	Washington (1)
Elk (1)	Mitchell (1)	Wilson (2)
Ellis (1)	Ness (1)	Woodson (2)
Ellsworth (3)	Norton (1)	Wyandotte (1)

STATEWIDE

Verdigris River, Spring River, and other streams and lakes throughout the state.

Pearl. (7)

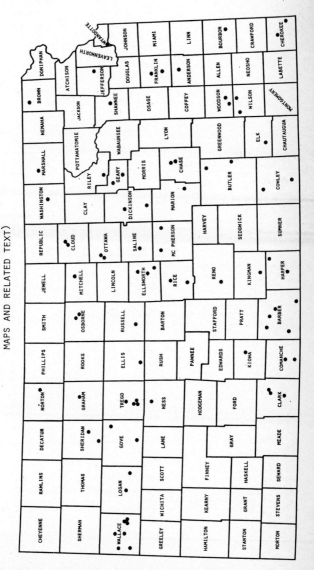

COLLECTING SITES IN KANSAS

(FOR SPECIFIC DETAILS ON THESE LOCATIONS, SEE THE FOLLOWING MAPS AND RELATED TEXT)

ANDERSON COUNTY

1. Garnett area: 3 miles north; at the Railroad Quarry.
   Fossils: flora in wide variety; to 6" long. (5-sandstone
   -12)

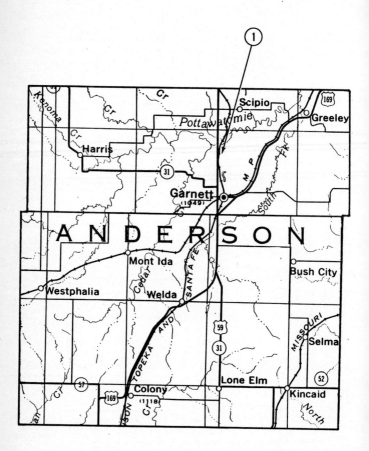

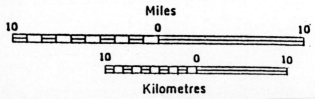

BARBER COUNTY

1. Aetna area.
   Agate. Jasper. Petrified wood. (1-13)

2. Kiowa area: along the Medicine Lodge River to Gerlane.
   Agate. Jasper. (7-20)

3. Lake City area: upstream, in and along the Medicine Lodge
   River all the way to the Kiowa County border; in
   streambed, on both banks, and on adjacent buttes and
   terraces.
   Agate. Jasper. Petrified wood. (1-3-7-13-20)

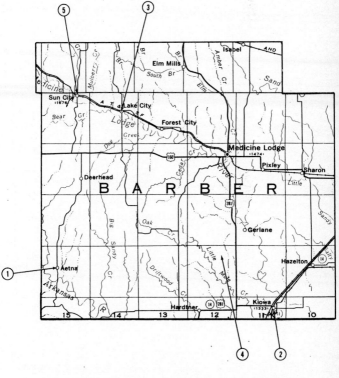

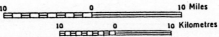

4. Medicine Lodge area: 10 miles south on US-281; 1.6 miles southwest.

    Petrified wood. (1-13)

5. Sun City area: at the gypsum works.

    Calcite. Gypsum. (2)

BOURBON COUNTY

1. Fort Scott area: in Pennsylvanian coal measures of the Marais des Cygnes.

    Siderite. (5-coal -39)

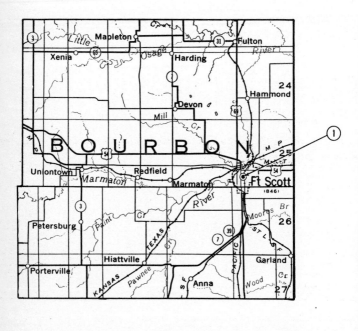

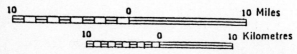

BROWN COUNTY

1. Morrill area: 2.4 miles northwest; in and adjacent to Pony
   Creek.

      Celestite: pink; crystals in clusters. (1-7-13-14-19-20-
      28)

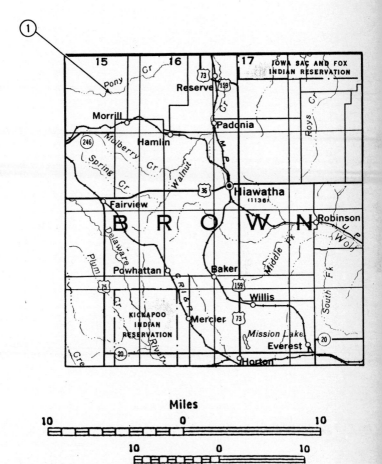

Miles

Kilometres

BUTLER COUNTY

1. Cassoday area: from headwaters of the Walnut River;
   downstream along both banks and in adjacent hills, all the
   way southwest to the Cowley County border.

   Calcite: geodes to 7″ diameter; well formed; nicely
   crystal-lined. Quartz: geodes to 7″ diameter; well
   formed; nicely crystal-lined. (1-7-13-20)

2. Rosalia area: 3.8 miles east on US-54; in road cut near the
   Greenwood County border.

   Quartz: geodes to 12″ diameter. (4)

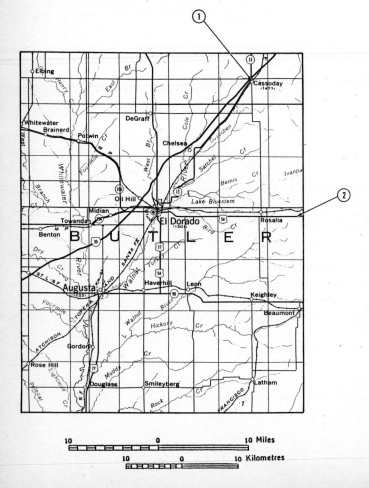

CHASE COUNTY

1. Cottonwood Falls area.

    Fossils: primarily marine. (5-limestone -12)

2. Strong City area: upstream and downstream for several miles in both directions along Cottonwood Creek.

    Calcite: geodes to 7″ diameter; well formed; crystal-lined. Quartz: geodes to 7″ diameter; well formed; crystal-lined. (1-7-13-20)

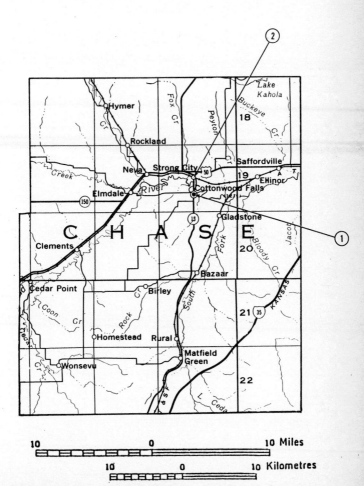

CHEROKEE COUNTY

1. County-wide in coal strip mines.

   Calcite: fine crystals. Dolomite. Marcasite. Pyrite: excellent brassy cubic crystals. (2)

2. Galena area mines.

   Calcite: fine; honey-colored crystals. Chalcopyrite: iridescent crystals. Dolomite: deep pink crystals. Galena: brilliant cubic crystals. Marcasite: sparkling cockscomb. Sphalerite: deep ruby-red crystals; many with iridescent blue-green planes. (2)

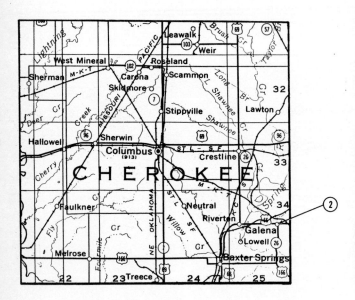

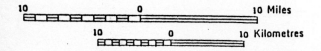

CLARK COUNTY

1. Ashland area: north several miles; near Mount Casino.
   Agate, moss. (3)

2. Ashland area: 9.2 miles north on SR-94 to Bluff Creek; on hill beside creek.
   Opal, moss. (1-13)

3. Ashland area: 10 miles north on SR-94 to Bluff Creek; upstream and downstream; especially at base of steep bluffs.
   Opal, moss. (1-13-14-20)

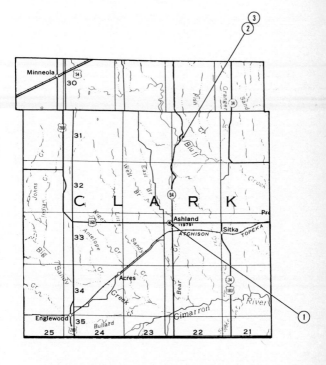

CLOUD COUNTY

1. Concordia area: just north; in and along the area of the Republican River called the Old Stream Bed (aka Old River).

    Agate, banded. Agate, common: clear. Agate, moss. Jasper. Petrified wood. (1-3-13-20)

2. Concordia area: in gravel pits adjacent to the Republican River.

    Agate. Chalcedony. (3)

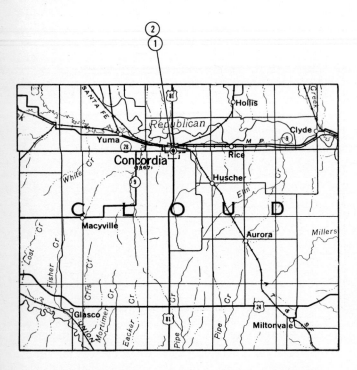

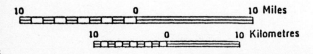

COMANCHE COUNTY

1. County-wide in mines and quarries.
   Calcite: good crystals. Gypsum: good crystals. (2-12)

2. County-wide in southern part of the county generally, but especially in the Salt Fork Creek valley.
   Gypsum: crystals. (1-13-39)

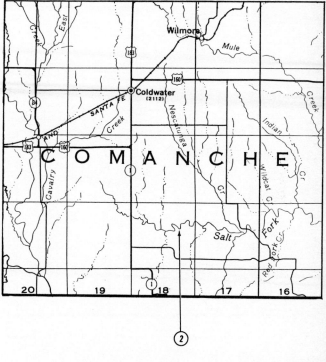

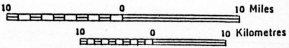

COWLEY COUNTY

1. Rock area: in and along the Walnut River from the Butler
   County border downstream on both sides and in adjacent
   hills, south to the Oklahoma border.

    Calcite: geodes to 7″ diameter; well formed; nicely
    crystal-lined. Quartz: geodes to 7″ diameter; well
    formed; nicely crystal-lined. (1-7-13-20)

2. Silverdale area.

    Fossils: primarily marine. (5-limestone -12)

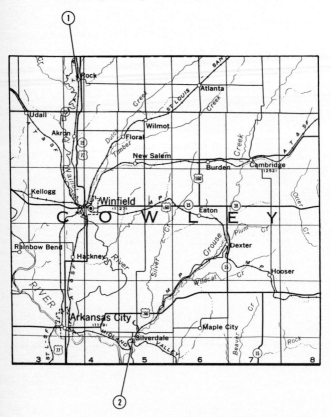

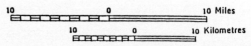

10   0   10 Miles

10   0   10 Kilometres

DICKINSON COUNTY

1. Abilene area: 2 miles south on SR-15 to Smoky Hill River;
   collect on both sides of the river upstream to the Saline
   County border and downstream to Chapman.

   Opal, moss. (1-3-7-13-20)

2. Elmo area: 3.5 miles southeast.

> Fossils: flora and fauna in wide variety (at this site was found a superb fossil dragonfly with 30″ wingspread). (28-shale -29)

ELK COUNTY

1. Moline area.

> Fossils: primarily marine; especially fine cephalopods. (5-limestone -12)

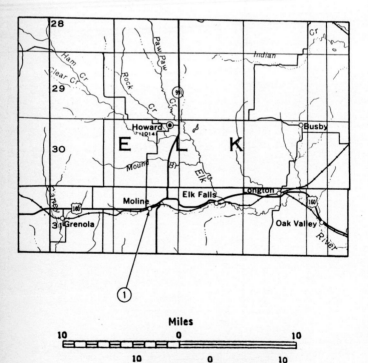

Miles

Kilometres

ELLIS COUNTY

1. Schoenchen area: along both banks and on tops and sides of hills and bluffs flanking the Smoky Hill river upstream to the Trego County border in the west and downstream to the Russell County border in the east.

    Agate. Petrified wood. (1-3-7-13-20)

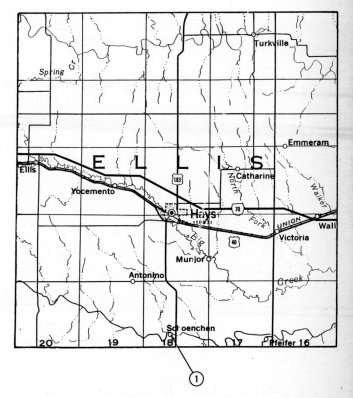

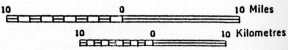

ELLSWORTH COUNTY

1. Carneiro area: at the nearby location called Mushroom Rocks.

    Pyrite: to 6″ diameter. (1-13-41)

2. Carneiro area: 3.75 miles south; just north of Kanopolis Lake.

    Agate. Fossils: silicified bone. Jasper. Petrified wood. (3)

3. Langley area: 3.75 miles north on SR-141 to Lake Kanopolis; all around shoreline and surrounding area of lake; celestite crystals especially in area of the dam.

    Barite: roses. Celestite: good crystals. (1-13-20)

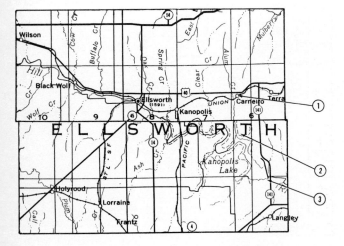

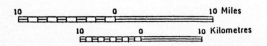

FRANKLIN COUNTY

1. Ottawa area: at the Rose Quarry.

    Fossils: flora in wide variety: leaves, stalks, cones, seeds, pods, club mosses, horsetails; to 6″ long. (5-sandstone -12)

2. Ottawa area: 4 miles south on US-59; at the Buildex Quarry.

    Fossils: flora in wide variety: leaves, stalks, cones, seeds, pods, club mosses, horsetails; to 6″ long. (5-sandstone -12)

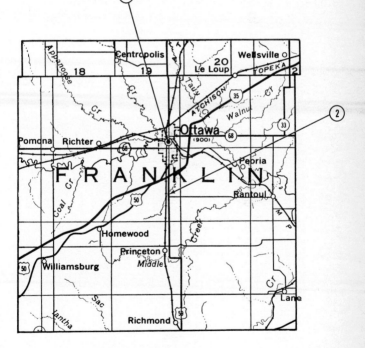

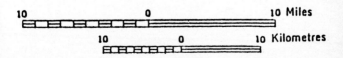

GEARY COUNTY

1. County-wide.

    Calcite: geodes to 6″ diameter. Petrified wood. Quartz: geodes to 6″ diameter. (4-10-11)

2. Junction City area: north along the Republican River to the Milford Lake Dam.

    Agate. Jasper. (1-3-7-13-20)

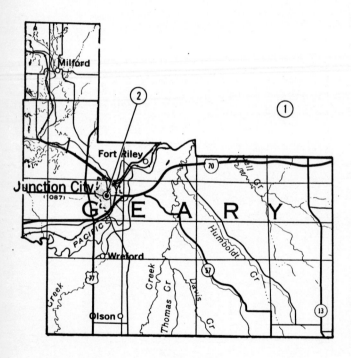

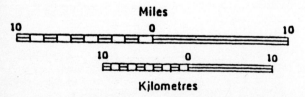

GOVE COUNTY

1. Gove area: 13.75 miles south on SR-23 to bridge over Smoky Hill River; collect upstream on both sides and on sides and tops of flanking hills and bluffs to the Logan County border and downstream to Trego County border.

    Agate. Chalcedony. Petrified wood. (1-3-5-chalk bluffs -7-13-14-20)

2. Quinter area.

    Calcite: to 16″ diameter. (1-13-31-volcanic -41)

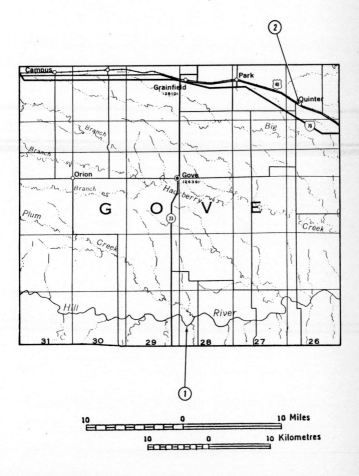

GRAHAM COUNTY

1. Hill City area: west; in road cuts along US-24 to Hoxie (in Sheridan County).

   Opal, moss. (5-limestone -14-29)

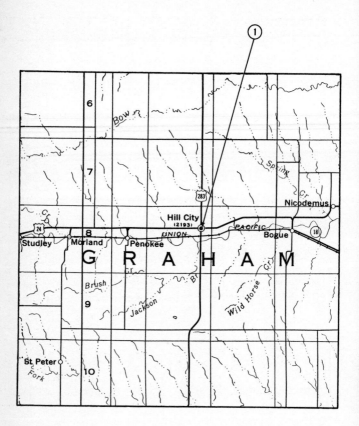

HARPER COUNTY

1. Anthony area salt mines.
   Halite: good crystals. (2)

2. Runnymede area: 1.4 miles north on SR-2 to bridge over the Chikaskia River; collect upstream along river to the Kingman County border and downstream to the Sumner County border.
   Agate. Jasper. Petrified wood. (3-18)

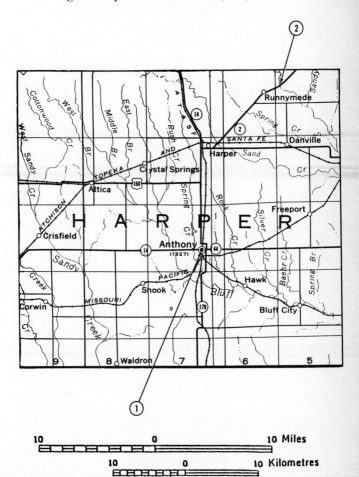

JEFFERSON COUNTY

1. McLouth area: in glacial drift.

   Agate, fortification (Lake Superior var.). Chalcedony.
   Jasper. Petrified wood. (3-10)

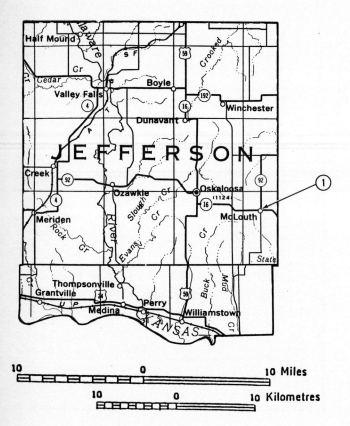

KINGMAN COUNTY

1. Rago area: upstream along both sides of the Chikaskia
   River to the Harper County border.

   Agate. Jasper. Petrified wood. (1-3-7-13-20)

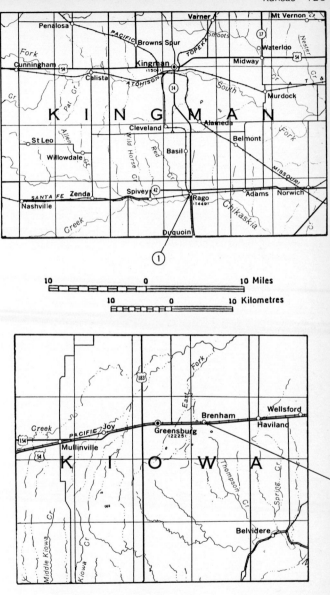

10   0   10 Miles

10   0   10 Kilometres

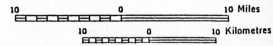

10   0   10 Miles

10   0   10 Kilometres

KIOWA COUNTY

1. Brenham area: in 5-mile radius; especially at Kimberly Ranch (often found with metal detectors).

    Meteorites (aka Pallasites): stony iron var. (1-6-13)

LOGAN COUNTY

1. Elkader area: west.
    Agate. (1-7-20)

2. Russell Springs area: upstream on both sides of Smoky Hill River and on tops and sides of hills and bluffs flanking the river upstream to the Wallace County border and downstream to the Gove County border.

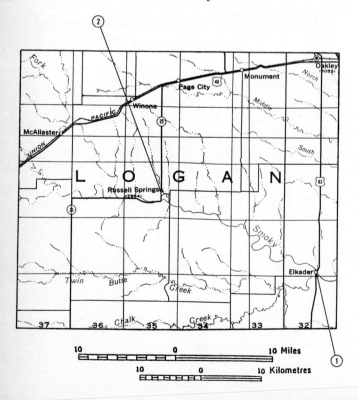

Agate. Chalcedony. Fossils: flora; many silicified (opalized). Opal, common: clear to white; manganese dendrites. Petrified wood. (1-3-5-chalk bluffs -7-13-14-20)

MARION COUNTY

1. Florence area: especially upstream and down in Cottonwood Creek.

Calcite: geodes to 7″ diameter; well formed; crystal-lined. Quartz: geodes to 7″ diameter; well formed; crystal-lined. (1-7-13-20)

MARSHALL COUNTY

1. Marysville area: upstream on both sides of Big Blue River
   and on flanking hills and bluffs to Nebraska border and
   downstream to Tuttle Creek Lake, 5 miles south of the
   US-77 bridge.

   Agate. Chalcedony: pebbles. (1-7-13-20)

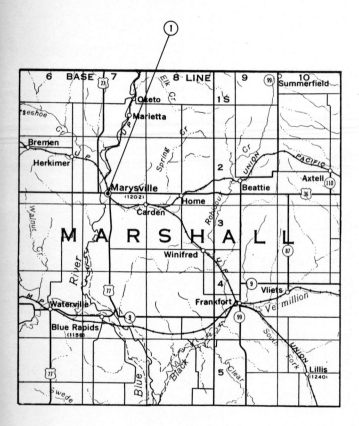

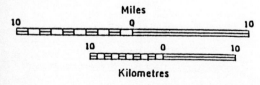

McPHERSON COUNTY

1. Freemount area: south of SR-4.

    Septarian: concretions to 25″ diameter; finest quality; yellow calcite seams; many with cavities containing perfectly terminated bright yellow calcite crystals or drusy. (1-13-31-41)

Miles

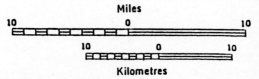

Kilometres

2. Roxbury area: at Roxbury Hill.

    Septarian: concretions to 25″ diameter; finest quality; yellow calcite seams; many with cavities containing perfectly terminated bright yellow calcite crystals or drusy. (1-4-7-13-20-31-41)

MITCHELL COUNTY

1. Beloit area: west; in the Blue Hills.

    Septarian: nodules to 16″ diameter. (1-7-13-14-20-29)

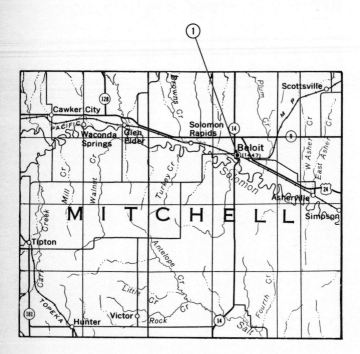

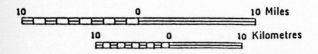

NESS COUNTY

1. Ransom area: 1.75 miles east on SR-4 to left turn on US-283; in road cuts north on US-283 to Trego County border.

    Agate. Petrified wood. (4)

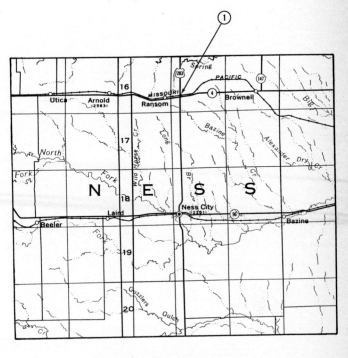

NORTON COUNTY

1. Calvert area.

    Septarian: concretions to 16″ diameter; calcite seams and crystal formations in cavities. (1-13-20-31-41)

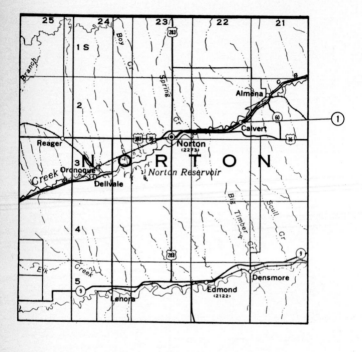

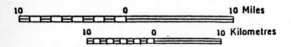

OSBORNE COUNTY

1. Osborne area: 0.5 mile south of Hobbie Lake.

    Septarian: concretions to 30″ diameter; fine calcite in bright yellow seams and well-terminated crystals in cavities; some cavities lined with calcite drusy. (1-13-20-31-41)

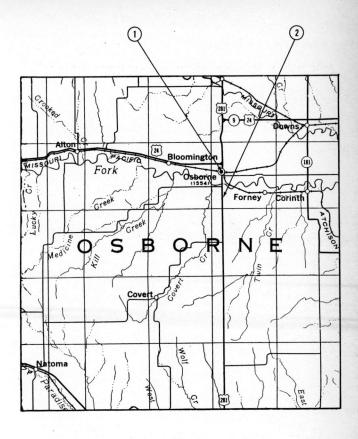

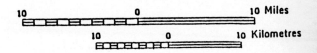

2. Osborne area: 2 miles south on US-281.

Septarian: concretions to 30″ diameter; fine calcite in bright yellow seams and well-terminated crystals in cavities; some cavities lined with calcite drusy. (1-13-20-31-41)

OTTAWA COUNTY

1. Ada area: 1 mile north.
   Petrified wood. (1-13)

2. Ada area: 1.5 miles northeast.
   Petrified wood. (1-13)

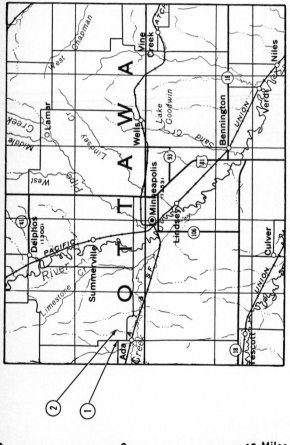

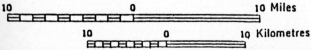

RENO COUNTY

1. Arlington area salt mines.
   Halite: large fine crystals. (2)

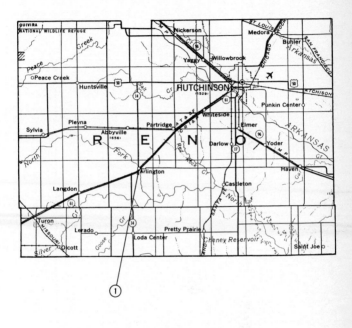

RICE COUNTY

1. Lyons area salt mines.
   Halite: good crystals. (2)

2. Sterling area salt mines.
   Halite: good crystals. (2)

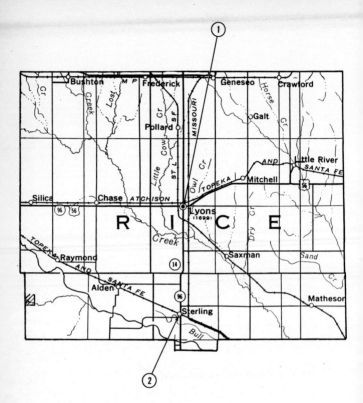

RILEY COUNTY

1. Ogden area.

   Garnet, pyrope: fine gem quality. (5-29-igneous rock)

Miles

Kilometres

RUSSELL COUNTY

1. Russell area: 7.2 miles south on US-281 to bridge over the Smoky Hill River; collect on both sides upstream to the Ellis County border.

   Agate. Petrified wood. (1-3-7-13-20)

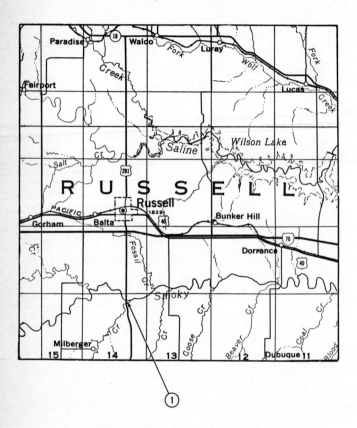

SALINE COUNTY

1. Bavaria area: 2 miles east; at the old Saline Quarry.
    Barite: roses; pale tan. (12)

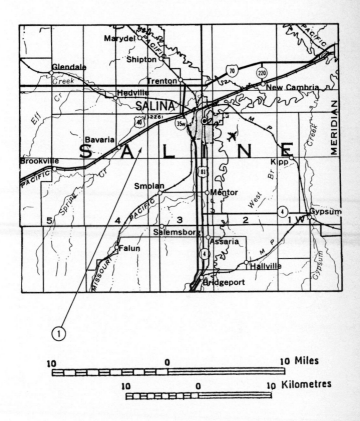

SHAWNEE COUNTY

1. Topeka area: in glacial moraines.
    Agate, banded. (3)

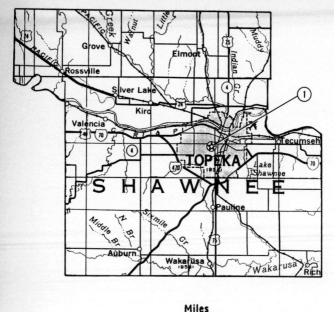

Miles

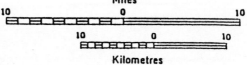

Kilometres

SHERIDAN COUNTY

1. Hoxie area: 12.2 miles south on SR-23 to bridge over the
   Saline River; collect downstream to southeast corner of
   this county and upstream to Thomas County border.

   Opal, moss. (5-limestone -14-29)

2. Tasco area: collect upstream (south-southwest) along
   South Fork of Solomon River all the way to the Thomas
   County border.

   Opal, moss. (5-limestone -14-29)

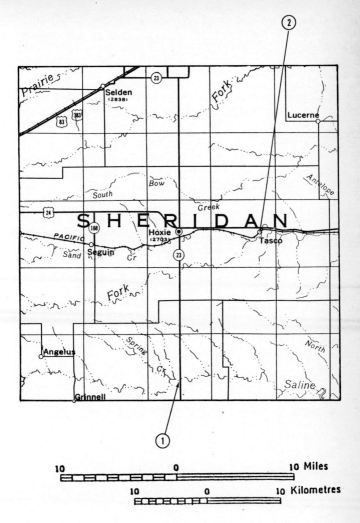

10   0   10 Miles

10   0   10 Kilometres

TREGO COUNTY

1. County-wide; in valleys.

     Calcite: geodes to 8″ diameter. Chalcedony: geodes to 8″ diameter. Quartz: geodes to 8″ diameter. (1-13)

2. Trego Center area: 6.25 miles south on US-283 to Smoky Hill River; collect along both sides and on flanking hills and bluffs upstream (west) to the Gove County border.

   Agate. Chalcedony. (1-3)

3. Trego Center area: 6.25 miles south on US-283 to Smoky Hill River bridge; then 2.9 miles east (downstream) to Cedar Bluff Reservoir; on shores (especially south shore) of reservoir.

   Fossils: flora; many silicified (opalized). Opal, common: clear to opaque white; manganese dendrites. (5-chalk bluffs -14)

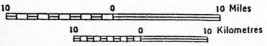

4. Trego Center area: 6.75 miles south on US-283 to just south of Smoky Hill River bridge; at the Seybert Sand Pit.

 Agate. Petrified wood. (11)

WALLACE COUNTY

1. County-wide; especially in valleys.

 Calcite: geodes to 8″ diameter. Chalcedony: geodes to 8″ diameter. Quartz: geodes to 8″ diameter. (1-13)

2. Sharon Springs area: north to sand pit.

 Agate. Petrified wood. (11)

3. Wallace area: 1.5 miles south to Smoky Hill River, south branch; collect along both sides and on tops and sides of hills and bluffs flanking river upstream (west) to the Colorado border and downstream (east) to the Logan County border.

 Agate. Petrified wood. (1-7-13-20)

4. Wallace area: 1.5 miles south on county road to Smoky Hill River bridge; then 0.25 mile upstream; from south shore of river to approximately 500 yards south of river.

 Opal, cachalong: green fire. (1)

5. Wallace area: 1.5 miles south on county road to Smoky Hill River valley; in road cuts north and south of bridge.

 Opal, common: clear to opaque white; manganese dendrites; all weathered specimens have dull white coating. (4)

6. Weskan area: 3.9 miles south to bridge over Ladder Creek; collect on both sides, upstream to Colorado border and downstream to Greeley County border.

 Opal, common: clear to opaque white; manganese dendrites. (7-20)

7. Weskan area: 4.1 miles north to bridge crossing Willow Creek; collect in streambed and on both sides from mouth at Smoky Hill River; 4.8 miles downstream (southeast) of bridge all the way upstream (northwest) to the Colorado border.

> Opal, common: clear to opaque white; manganese dendrites. (1-7-13-20)

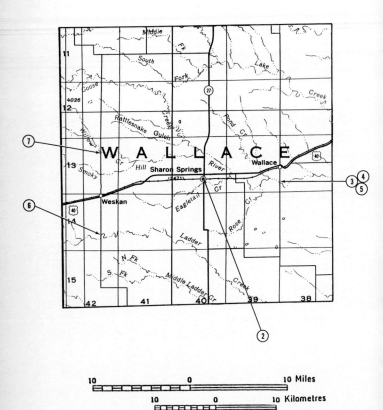

WASHINGTON COUNTY

1. Washington area: upstream and down on Mill Creek.
   Celestite: good crystals. (1-7-13-20)

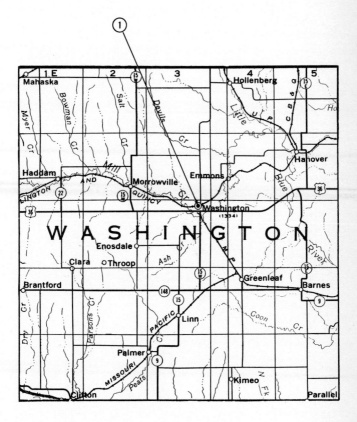

WILSON COUNTY

1. Buffalo area: upstream and down along East Buffalo Creek.

    Agate. (7-20)

2. Coyville area: downstream on the Verdigris River from the Toronto Lake Dam to just below Benedict.

    Agate. Chalcedony. Jasper. Petrified wood. (1-3-7-20)

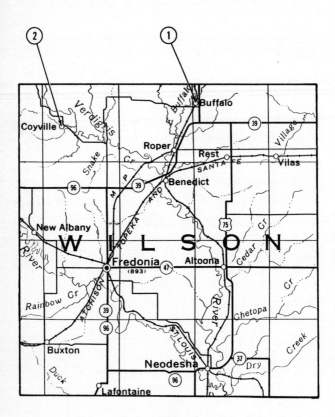

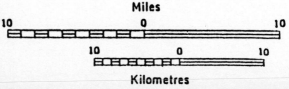

WOODSON COUNTY

1. Rose area: 1.9 miles west.
     Quartz, rock. (5-granite)

2. Rose area: 5 miles west-southwest.
     Amethyst. (5-granite)

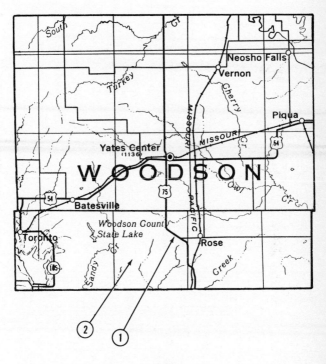

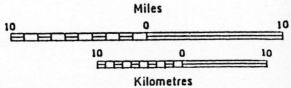

Miles

Kilometres

WYANDOTTE COUNTY

1. Bonner Springs area: in the Lone Star Quarry.
   Quartz: geodes to 9″ diameter. (12)

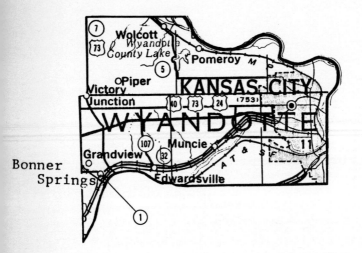

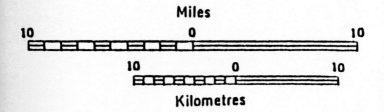

MINNESOTA

Statewide total of 82 locations in the following counties:

Blue Earth (1)	Fillmore (1)	Pine (1)
Carlton (7)	Goodhue (2)	Ramsey (1)
Cass (1)	Hennepin (2)	St. Louis (17)
Chippewa (1)	Houston (1)	Scott (1)
Chisago (1)	Itasca (2)	Swift (1)
Cook (9)	Lake (5)	Wabasha (3)
Crow Wing (8)	Le Sueur (1)	Washington (3)
Dakota (1)	Morrison (6)	Winona (2)
Faribault (3)	Olmsted (1)	

STATEWIDE

St. Peters River and other streams; Lake-of-the-Woods and other lakes.

Pearl. (7)

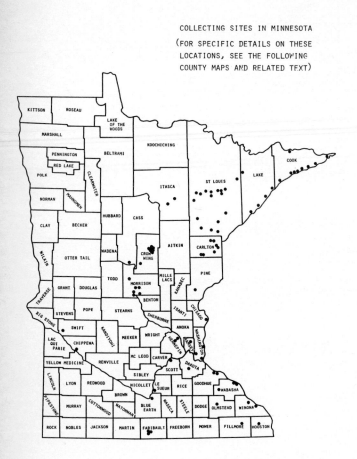

COLLECTING SITES IN MINNESOTA

(FOR SPECIFIC DETAILS ON THESE
LOCATIONS, SEE THE FOLLOWING
COUNTY MAPS AND RELATED TEXT)

BLUE EARTH COUNTY

1. Mankato area gravel pits.
    Agate, fortification. (10)

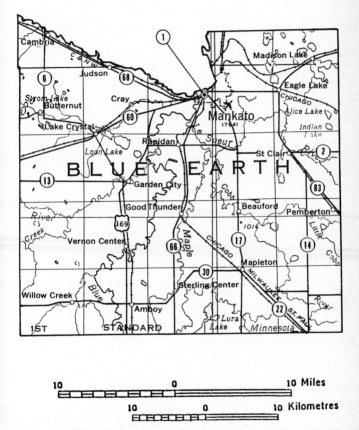

CARLTON COUNTY

1. Blackhoof area: in Twin Lakes Township quarry.
    Agate, fortification (Lake Superior var.): some very large (including largest ever recorded, a 108-lb nodule found in 1979). (10)

2. Carlton area mine dumps.

> Agate, fortification. Garnet. Greenalite. Magnetite.
> Marcasite. Minnesotaite. Pyrite. (2)

3. Carlton area: in the Carter Gravel Pit.

> Agate, fortification. (10)

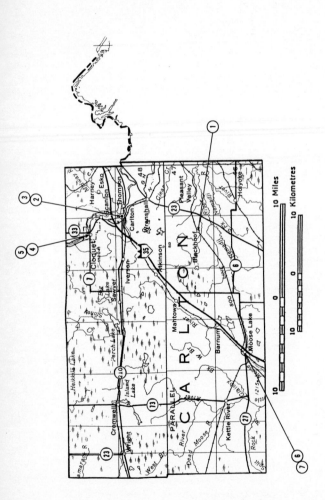

4. Cloquet area gravel pits and excavations.

> Agate, fortification (Lake Superior var.): some very large (including nodule of 24 lb found in excavation near the Cloquet Armory). (10)

5. Cloquet area mine dumps.

> Agate, fortification. Garnet. Greenalite. Magnetite. Marcasite. Minnesotaite. Pyrite. (2)

6. Moose Lake area: along the shores of Moose Lake and other lakes and streams nearby.

> Agate, fortification (Lake Superior var.): very fine; some quite large (including 49.75-lb nodule found in 1976). (20)

7. Moose Lake area mine dumps.

> Agate, fortification. Garnet. Greenalite. Magnetite. Marcasite. Minnesotaite. Pyrite. (2)

CASS COUNTY

1. Pillager area.

> Agate, fortification (Lake Superior var.): fine; some very large (including 4.5-lb nodule found in 1964). (1-3-10-13)

CHIPPEWA COUNTY

1. Montevideo area gravel pits.

> Agate, fortification. (10)

CHISAGO COUNTY

1. Taylors Falls area mines.

> Chalcocite, argentiferous. (2)

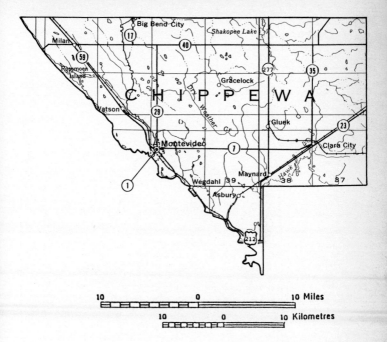

COOK COUNTY

1. Grand Marais area: at Good Harbor Bay on Lake Superior shoreline.

    Agate, fortification. Lintonite. Thomsonite. (20)

2. Grand Marais area: 3.9 miles northeast on US-61 to Devil Track River; cross bridge and continue on US-61 another 1.5 miles to side road left; follow this to pistol range; then follow trace road northward for 100 yards to agate bed.

    Agate, fortification: nodules to 5″ diameter. (1)

3. Grand Marais area: 3 miles southwest on US-61; in upper ledge of road cut where power line crosses.

    Agate, fortification. Thomsonite. (4)

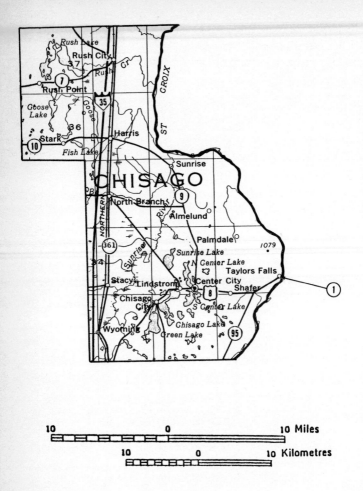

4. Grand Marais area: 4.5 miles southwest on US-61; collect along Thomsonite Beach for 0.25 mile southward to Terrace Point.

> Thomsonite: finest and most abundant in U.S.; usually round; beautifully "eyed"; exterior ordinarily black or dark green but interior is green, pink, or chartreuse. (5-basalt -20)

5. Grand Marais area: 13 miles northeast on US-61; 1.2 miles southwest of the mouth of the Brule River; at Paradise Beach.

Agate, fortification. (5-basalt -20)

6. Grand Portage area: 1.2 miles south; on Lake Superior shoreline in area of the point.

Agate, fortification. (20)

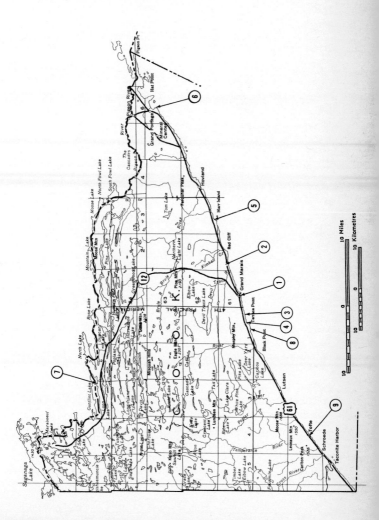

7. Gunflint Lake area: entire south shoreline.

    Agate, banded; superior quality. Agate, fortification: nodules to 8″ diameter; excellent color zonings. (3-5-7-20)

8. Lutsen area: northeast on US-61 to Thomsonite Beach area: at four small open-pit mines just west off highway.

    Agate, fortification (Lake Superior var.). Thomsonite. (2)

9. Tofte area: at Two Islands Beach.

    Agate, fortification. (7-20)

CROW WING COUNTY

1. Brainerd area gravel pits.

    Agate, fortification. (10)

2. Crosby area: at the Arco Mine dump.

    Agate, fortification. Binghamite (Silkstone). (2)

3. Crosby area: at the Hanna Company Mines Dump Number Two.

    Binghamite (Silkstone). (2)

4. Crosby area: north on SR-6 a short distance to the Evergreen Mine (aka Portsmouth Mine) dump; turn left on side road that bisects dump; collect on both sides of road.

    Agate, fortification. Binghamite (Silkstone). (2)

5. Crosby area: north on SR-6 a short distance to the Evergreen Mine (aka Portsmouth Mine) dump; turn left on side road that bisects dump; continue west on this road to where it turns south; collect on left at the Louis Mine dump.

    Binghamite (Silkstone). (2)

6. Crosby area: southwest on SR-210 to the Hopkins Mine dump.

    Rhodochrosite. (2)

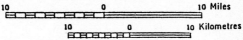

10   0   10 Miles

10   0   10 Kilometres

7. Crosby area: southwest on SR-210 to the Pittsburgh-Pacific Mine.

    Rhodochrosite. (2)

8. Ironton area: at the Maroco Mine.

    Pyrolusite. (2)

DAKOTA COUNTY

1. Farmington area: at gravel pit adjacent to Vermillion River.

    Agate, fortification. (10)

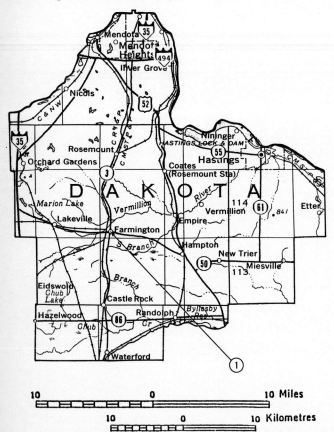

FARIBAULT COUNTY

1. Blue Earth area gravel pits.
     Agate, fortification (Lake Superior var.). (7-10-12-20)

2. Blue Earth area quarries.
     Agate, fortification (Lake Superior var.). (7-10-12-20)

3. Blue Earth area stream valleys.
     Agate, fortification (Lake Superior var.). (7-10-12-20)

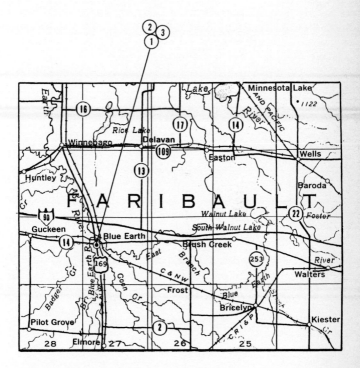

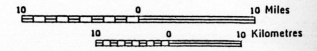

FILLMORE COUNTY

1. Lanesboro area: 1 mile northeast on SR-16; in road cut.
   Quartz, rock. (4)

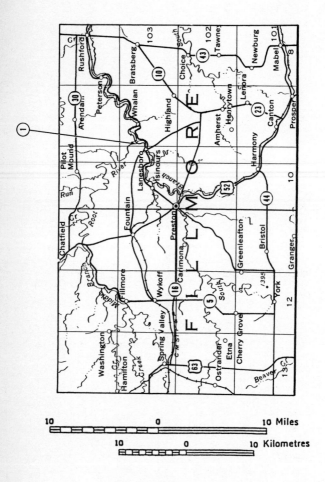

GOODHUE COUNTY

1. Red Wing area gravel pits.
   Agate, fortification. (10)

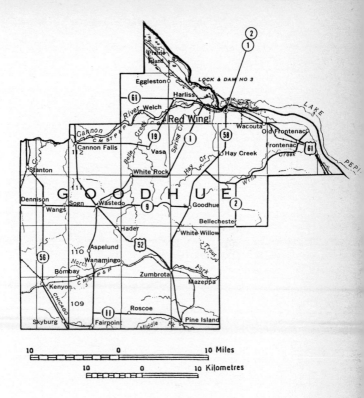

2. Red Wing area: in Mississippi River bluffs.
   Quartz, rock. (8-9-limestone -14-20)

HENNEPIN COUNTY

1. Osseo area: 0.25 mile south on CR-18 to side road left;
   follow side road east to end at three gravel pits.
   Agate, fortification. Chalcedony. Jasper. (10)

2. Osseo area: south on CR-18 to side road right; follow side
   road west to the Anderson Aggregates gravel pit.
   Agate, fortification. (10)

HOUSTON COUNTY

1. Caledonia area sand pit.
   Calcite: crystals. (11)

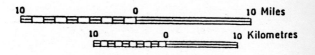

10   0   10 Miles

10   0   10 Kilometres

ITASCA COUNTY

1. Bovey area mine dumps.
    Agate, fortification. (10)

2. Grand Rapids area mine dumps.
    Agate, fortification. Garnet. Greenalite. Magnetite.
    Marcasite. Minnesotaite. Pyrite. (2)

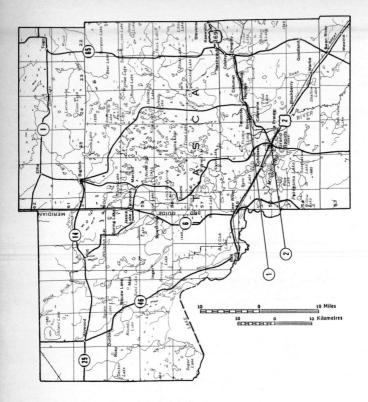

LAKE COUNTY

1. Beaver Bay area: on new gravel bars and beaches constantly being formed by wave action.

    Agate, fortification. (18-20)

2. Castle Danger area: 2.3 miles northeast on US-61; at Gooseberry Falls State Park beach.

    Agate, fortification. (20)

3. Illgen City area: 0.5 mile southwest on US-61 to mouth of Baptism River; park east of the Minnesota Highways Department blacktop storage depot and follow foot trail to lake; collect along shoreline south of big rock in water.

    Agate, fortification. (20)

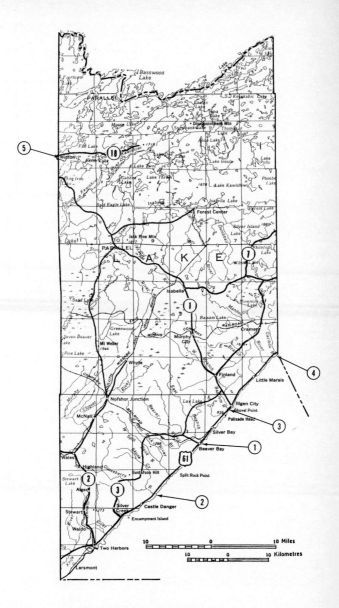

10 Miles

10 Kilometres

4. Little Marais area: 4.8 miles northeast on US-61; at County Line Beach.

   Agate, fortification. (20)

5. Winton area mine dumps.

   Jasper ("Evergreen" var.); green; gem quality. Quartz, rock. (2)

LE SUEUR COUNTY

1. Kasota area quarries along the Minnesota River.

   Fossils, marine: both vertebrate and invertebrate; many coral varieties. (12)

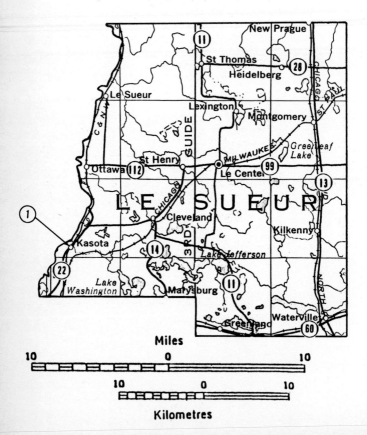

MORRISON COUNTY

1. Little Falls area: in gravels of Elk Creek.
    Agate, fortification. Garnet. Staurolite. (1-3-26-27)

2. Little Falls area gravel pits.
    Agate, fortification. (10)

3. Little Falls area: 6.3 miles south from edge of town on
    US-10; turn right (west) on paved township road and go 0.5
    mile; turn left (south) and go 1.6 miles to dirt road; follow
    dirt road 0.4 mile to parking area at Blanchard Dam;
    screen shoreline sands downstream on both sides of river
    below dam.
    Garnet. Staurolite. (1-3-7-20-26)

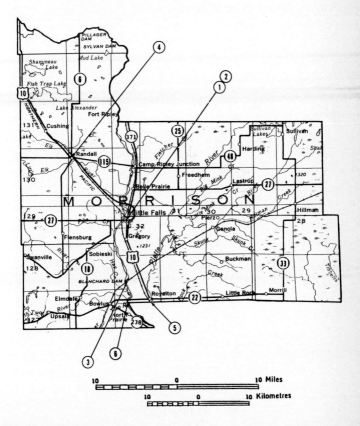

4. Randall area gravel pits.

   Agate, fortification. (10)

5. Royalton area: under Northern Pacific RR bridge over
   Platte River.

   Agate, fortification. Garnet. Staurolite. (3-27-29)

6. Royalton area: 3 miles west; along banks of Mississippi
   River.

   Garnet. Staurolite. (26-27)

OLMSTED COUNTY

1. Rochester area quarries and road cuts, especially to the
   east.

   Fossils, marine: primarily invertebrates. (4-10)

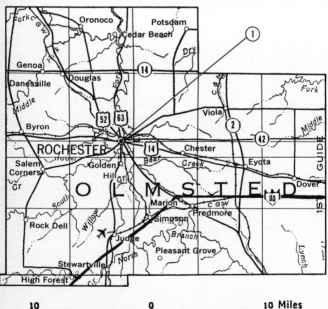

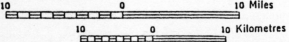

PINE COUNTY

1. Pine City area: along both banks of the Snake River.
   Chalcocite. (1-7-13-20-29)

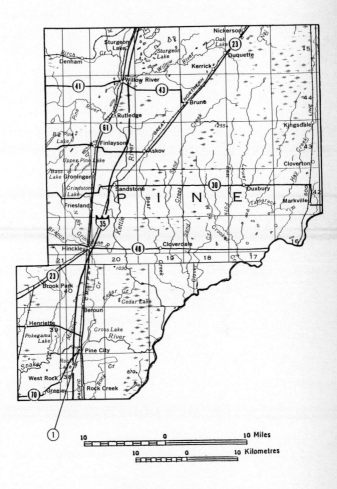

RAMSEY COUNTY

1. County-wide gravel pits.
   Agate, fortification. (10)

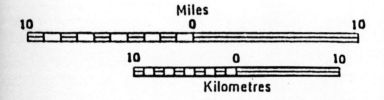

Miles

10   0   10

10   0   10

Kilometres

ST. LOUIS COUNTY

1. Aurora area mines and mine dumps.

   Agate, fortification. Fossils: algae; silicified. Jasper:
   fossil inclusions. (2)

2. Biwabik area: 1 mile west on SR-135 to side road marked
   with sign for the Mary Ellen Mine; take this road to the
   Corsica Mine dump.

   Jasper ("Mary Ellen" var.): banded. (2)

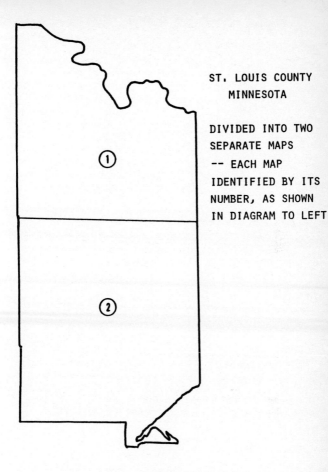

ST. LOUIS COUNTY
MINNESOTA

DIVIDED INTO TWO
SEPARATE MAPS
-- EACH MAP
IDENTIFIED BY ITS
NUMBER, AS SHOWN
IN DIAGRAM TO LEFT

3. Biwabik area: 1 mile west on SR-135 to side road marked with sign for the Mary Ellen Mine; take this road to the Mary Ellen Mine dump.

> Fossils: algae; silicified. Jasper ("Mary Ellen" var.): banded. (2)

4. Biwabik area: two miles west on SR-135; at the Corsica Mine.

> Fossils: algae; silicified. (2)

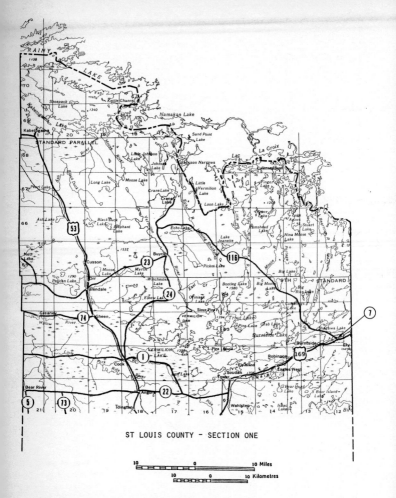

ST LOUIS COUNTY – SECTION ONE

5. Buhl area: east on US-169 to Top's Tavern; turn north on
   dirt road and go to Dormer Mine dumps on left.
   Agate, fortification. (2)

6. Chisholm area mines.
   Goethite: botryoidal; stalactic. (2)

7. Ely area mine dumps.

> Jasper ("Evergreen" var.); green; gem quality. Quartz, rock. (2)

8. Eveleth area mines and their dumps.

> Agate. Feldspar. Garnet. Greenalite. Magnetite. Marcasite. Minnesotaite. Pyrite. (2)

9. Fredenberg area: along shorelines of the Island Lake Reservoir.

> Agate, fortification. (20)

10. Floodwood area mine dumps.

> Agate, fortification. Garnet. Greenalite. Magnetite. Marcasite. Minnesotaite. Pyrite. (2)

11. Gowan area gravel pits.

> Agate, fortification. (10)

12. Hibbing area mines.

> Goethite: botryoidal; stalactic. (2)

13. Kinney area: at Wade Mine dumps.

> Agate, fortification. (2)

14. Munger area: at the Coon Brothers Gravel Pit.

> Agate, fortification. (10)

15. Saginaw area: 2.8 miles south on SR-33; at the Kinto Brothers Gravel Pit.

> Agate, fortification. (10)

16. Twig area: at the Arrowhead Gravel Pit.

> Agate, fortification. (10)

17. Virginia area.

> Agate, fortification (Lake Superior var.): fine; large (including 6-lb nodule found in 1966). (1-10-13)

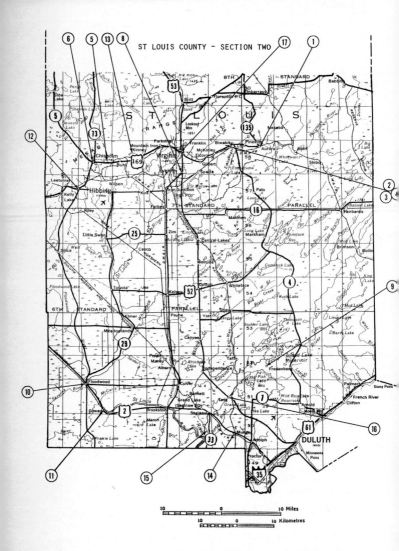

ST LOUIS COUNTY - SECTION TWO

SCOTT COUNTY

1. Shakopee area gravel pit.
   Agate, fortification. (10)

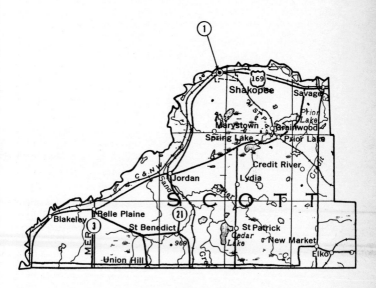

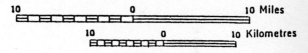

SWIFT COUNTY

1. Appleton area gravel pits.
   Agate, fortification. (10)

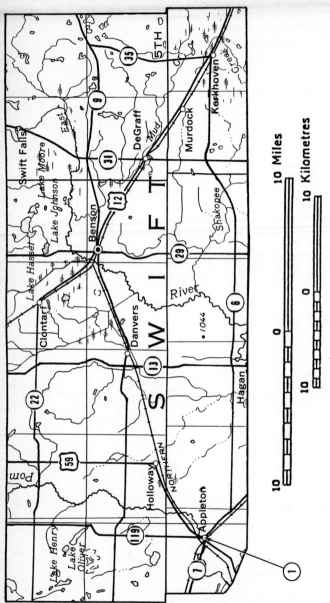

WABASHA COUNTY

1. Wabasha area gravel pits.

   Agate, fortification. (10)

2. Wabasha area: north edge of town; at the huge gravel pit near the RR station.

   Agate, fortification. (10)

3. Zumbro Falls area: collect upstream (west) along both sides of the Zumbro River and its tributaries to Mazeppa.

   Agate, fortification. (7-20)

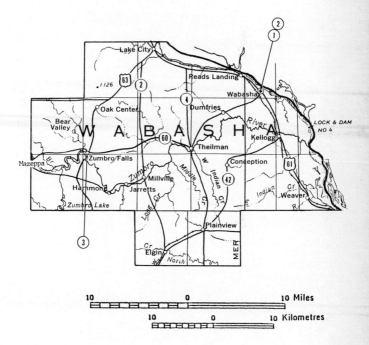

WASHINGTON COUNTY

1. Lakeland area: along west shoreline of Lake St. Croix.

   Agate, fortification. (7-20)

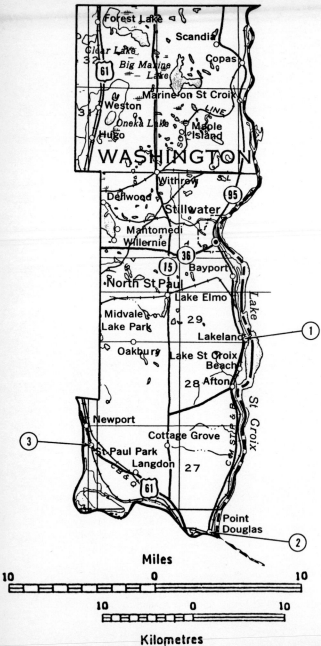

Miles

10   0   10

10   0   10

Kilometres

2. Point Douglas area: in gravels at mouth of the St. Croix
   River.
   Agate, fortification. (3-7-18-20)

3. St. Paul Park area gravel pit.
   Agate, fortification. Jasper: red, green. (10)

WINONA COUNTY

1. Goodview area: at large gravel pit on US-61.
   Agate, fortification. (10)

2. Winona area gravel pits.
   Agate, fortification; some of exceptional beauty. (10)

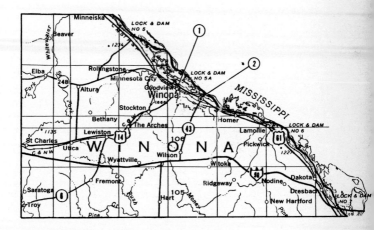

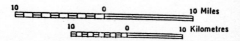

MISSOURI

Statewide total of 85 locations in the following counties:

Benton (4)	Franklin (1)	Miller (1)
Bollinger (1)	Gasconade (1)	Moniteau (1)
Boone (3)	Gentry (1)	Morgan (1)
Buchanan (2)	Greene (2)	Newton (4)
Callaway (3)	Grundy (1)	Ozark (1)
Cape Girardeau (1)	Henry (1)	Phelps (1)
	Hickory (1)	Polk (1)
Cass (3)	Iron (1)	Ralls (1)
Chariton (1)	Jackson (1)	Ripley (1)
Clark (8)	Jasper (2)	Ste. Genevieve (1)
Cole (3)	Jefferson (1)	
Cooper (1)	Lawrence (1)	St. Francois (1)
Crawford (2)	Lewis (1)	St. Louis (4)
Dade (1)	Livingston (1)	Shannon (3)
Daviess (2)	Madison (4)	Stoddard (2)
Dunklin (1)	McDonald (1)	Washington (5)

STATEWIDE

Current River, Jacks Ford, and other streams as well as some lakes.

Pearl. (7)

COLLECTING SITES IN MISSOURI

(FOR SPECIFIC DETAILS ON THESE
LOCATIONS, SEE THE FOLLOWING
COUNTY MAPS AND RELATED TEXT)

BENTON COUNTY

1. County-wide.

    Chalcedony. Chert (aka "Mozarkite"): ranging from
    pink to lavender blue; swirled patterns. Jasper: red,
    yellow. (1-3-13)

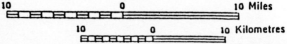

2. Lincoln area: at the Brice Patt farm (CF).

> Chert (aka "Mozarkite"): ranging from pink to lavender blue; swirled patterns. (1)

3. Lincoln area: in road cuts for 10 miles in all directions; especially in first road cut east of town on US-65.

> Chert (aka "Mozarkite"): ranging from pink to lavender blue; swirled patterns. (4)

4. Lincoln area: just south; downstream on both sides of tributary of Cole Camp Creek to its mouth at that larger creek.

> Chalcedony. Chert (aka "Mozarkite"): ranging from pink to lavender blue; swirled patterns. Fossils: marine. Jasper: red, yellow. (1-4-7-13-20)

BOLLINGER COUNTY

1. Lutesville area gravel pits, excavations, road cuts, RR cuts, and stream cuts eastward to the Cape Girardeau County border.

> Agate, fortification (Lake Superior var.). Petrified wood. (4-10-14-29)

BOONE COUNTY

1. Columbia area quarries and gravel pits.

> Agate: occasional. Chert. Fossils, marine: vertebrates and invertebrates. Petrified wood: waterworn; poor grade. (5-limestone -10-12)

2. County-wide.

> Fossils, marine: vertebrates and invertebrates. (5-limestone -12)

3. Huntsdale area: north shoreline of Missouri River downstream to Wilton.

> Agate: occasional. Chert. Fossils, marine. Petrified wood: pale coloration; poor grade. (7-13-20)

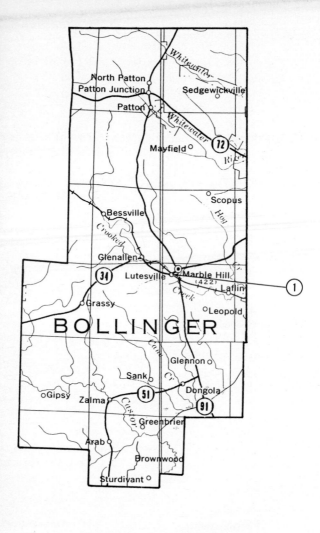

North Patton
Patton Junction
Sedgewickville
Patton
Whitewater River
72
Mayfield
Scopus
Bessville
Hoy
Crooked
Glenallen
34   Lutesville   Marble Hill (422)   Laflin   **1**
Creek
Grassy   Leopold
BOLLINGER
Cane Cr
Glennon
Sank   Cr
51   Dongola
Gipsy   Zalma   **91**
Castor   Greenbrier
Arab
Brownwood
Sturdivant

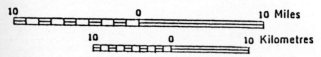

10   0   10 Miles

10   0   10 Kilometres

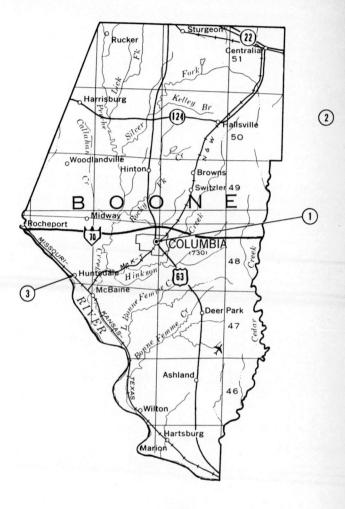

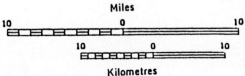

Miles

Kilometres

BUCHANAN COUNTY

1. St. Joseph area: at the Eighty-Nine Mine.

    Calcite: excellent crystals with flat terminations; transparent yellow, often with central area of crystal becoming translucent. (2)

2. St. Joseph area: at the Sweetwater Mine.

    Calcite: superb scalenohedron crystals; large; brilliant; chartreuse; distinct phantos. (2)

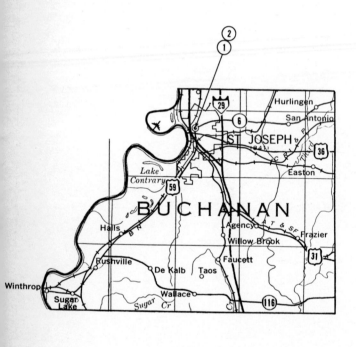

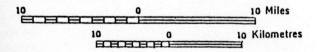

CALLAWAY COUNTY

1. County-wide.
   Fossils, marine: vertebrate and invertebrate. (5-limestone -12)

2. Fulton area quarries.
   Fossils, marine: vertebrate and invertebrate. (5-limestone -12)

3. New Bloomfield area: 2 miles southeast in tributary of Hillers Creek.

> Fossils, marine: vertebrate and invertebrate. (4-7-20)

CAPE GIRARDEAU COUNTY

1. Cape Girardeau area: extending outward virtually county-wide in all streambeds, excavations, road cuts, and RR cuts.

> Agate, fortification (Lake Superior var.). Petrified wood. (3-4-10-20)

CASS COUNTY

1. Archie area: 1.4 miles north on US-71; at quarry adjacent to South Grand River.

    Fossils, marine: vertebrate and invertebrate. Petrified wood. (12)

2. East Lynne area: just northeast in a quarry adjacent to Camp Branch.

    Chert. Fossils, marine: vertebrate and invertebrate. Petrified wood. (12)

3. Harrisonville area quarries.

    Petrified wood: light color but good quality. (12)

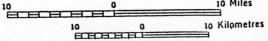

CHARITON COUNTY

1. Dalton area quarries.
    Tripoli. (12)

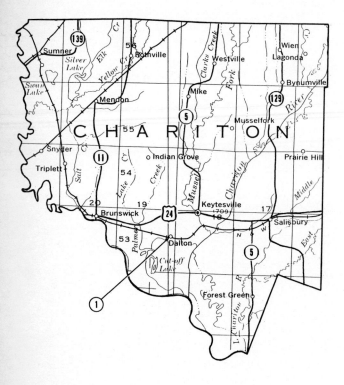

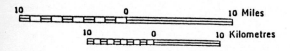

CLARK COUNTY

1. Alexandria area: 6 miles west on US-136 to left turn onto US-61; go a short distance south to gravel pit on the Scheffler property (CF).

    Quartz: geodes to 7″ diameter. (10)

2. Anson area: downstream on tributary of Des Moines River to its mouth.

    Quartz: geodes to 6″ diameter. (7-13-20)

3. Ashton area: 2.7 miles north on country road to bridge over South Branch of Fox River; collect upstream to Scotland County border and downstream to its mouth at Fox River.

    Quartz: geodes to 8″ diameter. (7-13-20)

4. Kahoka area: a short distance north on SR-81 to nursing home; go right 1 block and turn left; cross bridge over river and immediately turn right; turn left at next corner and follow to the Easterday farm (CF).

    Quartz: geodes to 6″ diameter. (1-7-13-20)

5. Kahoka area: 2.25 miles north on SR-81 to bridge over Fox River; collect upstream to Iowa border and downstream to Wayland area.

    Quartz: very abundant geodes to 10″ diameter; some with calcite inclusions. (1-7-13-14-20-28-29)

6. St. Francisville area: along south shore of the Des Moines River and especially up virtually all ravines, washes, gullies, and tributaries entering the river.

    Quartz: geodes to 11″ diameter. (1-3-7-13-14-20-26-28-29)

7. St. Francisville area: at the Sinotte property (CF).

    Jasper. Quartz: geodes to 6″ diameter. (1-7-13-20)

8. St. Francisville area: upstream in the Weaver's Branch.

    Chalcedony: geodes to 8″ diameter. Quartz: geodes to 8″ diameter; some with calcite inclusions. (1-3-7-13-18-20)

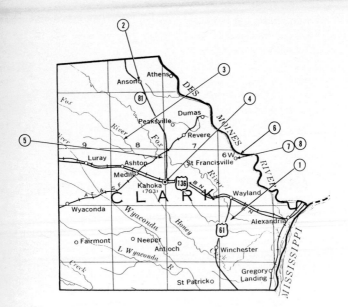

COLE COUNTY

1. Eugene area mines.
   Barite: crystals. (2)

2. Henley area mines.
   Barite: crystals. (2)

3. Hickory Hill area mines.
   Barite: crystals. (2)

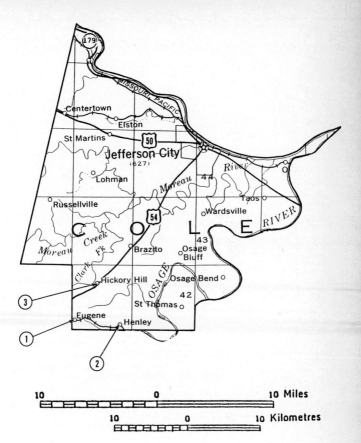

COOPER COUNTY

1. Clifton City area: at the old Sweeney Quarry.
   Fossils, marine: chiefly invertebrates. (12)

CRAWFORD COUNTY

1. Steelville area mines.
   Amethyst. Hematite. Pyrite. Quartz, rock. (2)

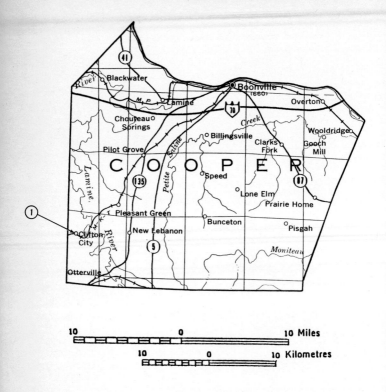

2. Steelville area: just east on SR-8 to mine dumps.
   Amethyst. Hematite. Pyrite. Quartz, rock. (2)

DADE COUNTY

1. County-wide in excavations, road cuts, RR cuts, and
   stream cuts on west slopes of the Ozark Mountains.
   Agate, fortification (Lake Superior var.). Chert: colorful; some gemmy. Petrified wood. (4-10-12-14-29)

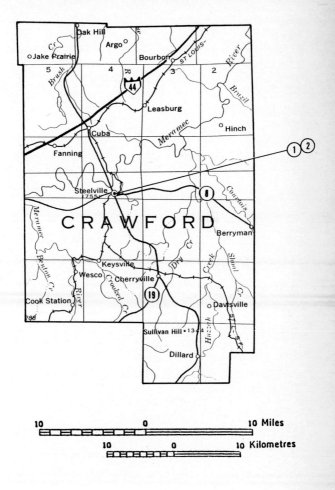

```
10          0          10 Miles
10          0          10 Kilometres
```

DAVIESS COUNTY

1. County-wide in gravel pits and in stream cuts, road cuts, RR cuts, and excavations.

    Agate, fortification (Lake Superior var.). Fossils, marine. Jasper. Petrified wood. (3-4-7-10-18-20)

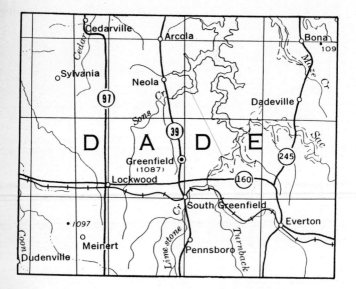

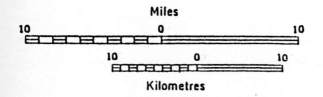

2. Gallatin area: upstream and down along both sides of the
   Grand River.

   Agate, fortification (Lake Superior var.). Chalcedony.
   Fossils, marine. Jasper. Petrified wood. (3-7-18-20)

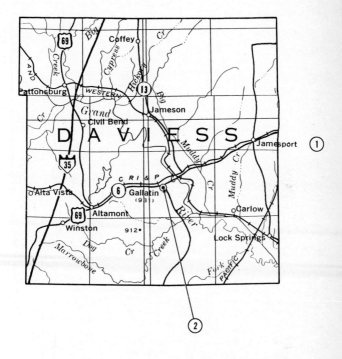

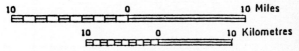

DUNKLIN COUNTY

1. Malden area: 0.8 mile west on CR-J to gravel road on the right; follow this to Cowley's Ridge; north to gravel pits.
   Chert. Jasper. (10)

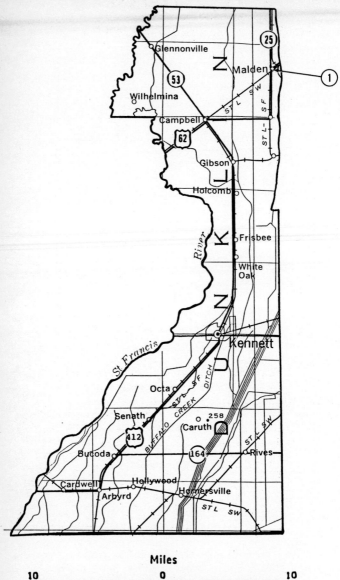

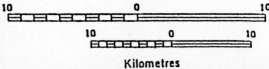

Miles

Kilometres

FRANKLIN COUNTY

1. Stanton area: at the Cherry Valley Iron Mine.
   Quartz, rock: rutilated. (2)

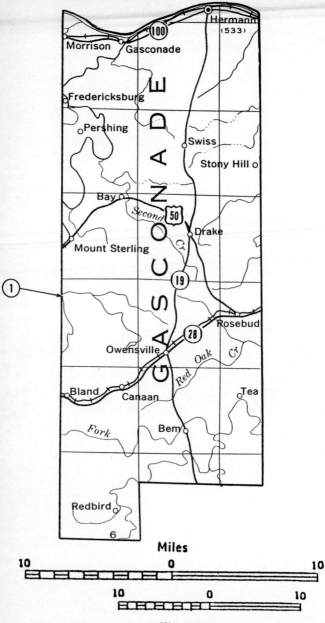

Miles

10   0   10

10   0   10

Kilometres

GASCONADE COUNTY

1. Owensville area: 8 miles west-northwest; in hills and abandoned clay pit.

    Fossils, marine: chiefly good brachiopods. (1-5-blue-gray quartzite -13-29)

GENTRY COUNTY

1. County-wide in gravel pits and excavations and in stream cuts, road cuts, and RR cuts.

    Agate, fortification (Lake Superior var.). Chalcedony. Fossils, marine. Jasper. Petrified wood. (3-4-7-10-18-20)

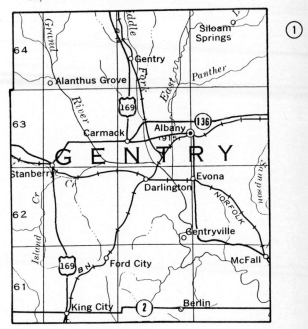

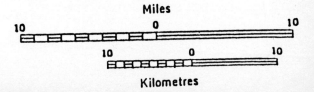

Miles

Kilometres

GREENE COUNTY

1. Springfield area: just north at Kings Butte.
   Fossils, marine: chiefly invertebrates. (1-13-14-29)

2. Springfield area quarries.
   Fossils, marine: chiefly invertebrates. (12)

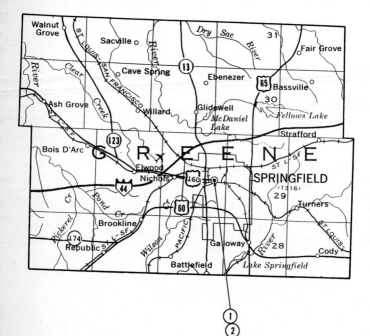

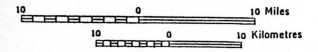

GRUNDY COUNTY

1. County-wide in gravel pits and excavations and in stream cuts, road cuts, and RR cuts.

   Agate, fortification (Lake Superior var.). Fossils, marine. Jasper. Petrified wood. (3-4-7-10-18-20)

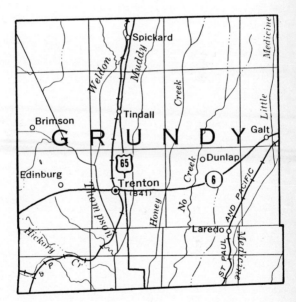

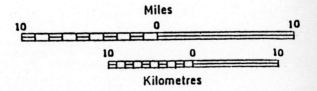

HENRY COUNTY

1. Urich area: 1.1 miles northwest on SR-7; 3 miles southeast
   of Creighton (Cass County); at quarry on the Cornett farm.
   Petrified wood. (12)

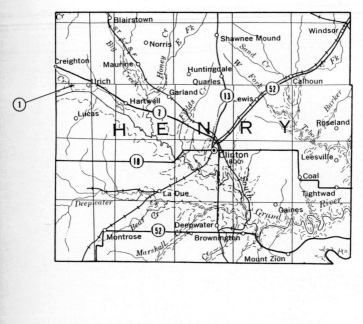

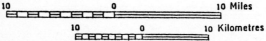

HICKORY COUNTY

1. Hermitage area.

   Chalcedony: mainly blue. Jasper: bluish, green, yellow. (1-3-4-13)

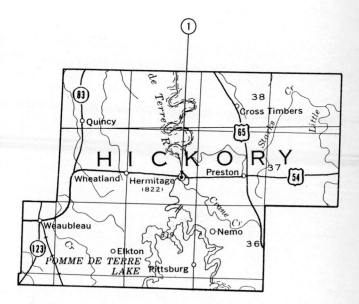

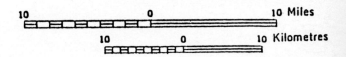

IRON COUNTY

1. Ironton area mines.
    Hematite: botryoidal clusters. (2)

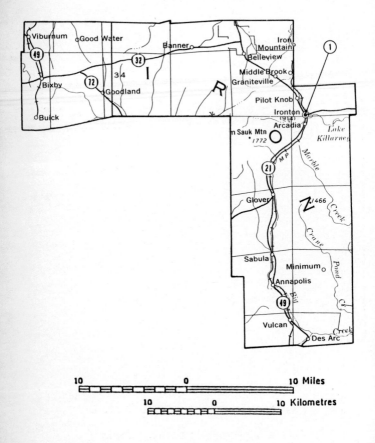

JACKSON COUNTY

1. Kansas City area.

　　Fossils, marine: vertebrates and invertebrates. (12)

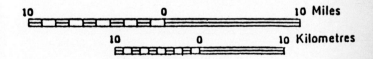

JASPER COUNTY

1. Carthage area quarries.

   Fossils, marine: chiefly invertebrates. (12)

2. Joplin area mines.

   Calcite: fine; honey-colored crystals. Chalcopyrite: iridescent crystals. Dolomite: deep pink crystals. Galena: brilliant cubic crystals. Marcasite: sparkling cockscomb. Sphalerite: deep ruby-red crystals, many with iridescent blue-green planes. (2)

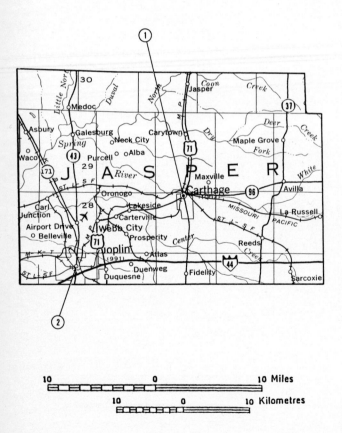

10   0   10 Miles

10   0   10 Kilometres

JEFFERSON COUNTY

1. De Soto area: southwest on SR-21 to CR-CC; follow it to CR-E; follow this road west to Tiff; then east over RR to mine.

    Agate. (2)

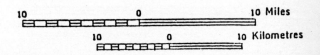

LAWRENCE COUNTY

1. Aurora area mines.
    Calamine. (2)

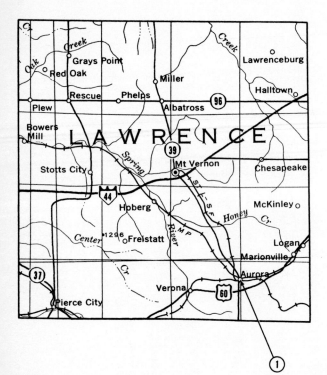

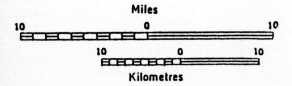

LEWIS COUNTY

1. La Grange area gravel-dredging operation.
   Agate: small but high quality. Jasper. Petrified wood.
   (10)

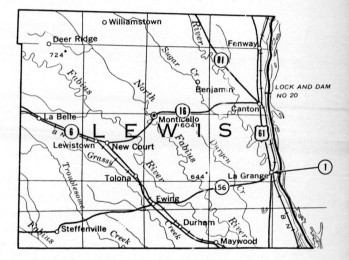

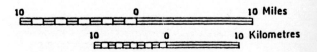

LIVINGSTON COUNTY

1. County-wide in gravel pits and excavations and in stream
   cuts, road cuts, and RR cuts.

   Agate, fortification (Lake Superior var.). Fossils, marine. Jasper. Petrified wood. (3-4-7-10-18-20)

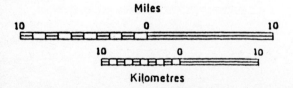

MADISON COUNTY

1. Fredericktown area mines.
    Chalcopyrite. (2)

2. Fredericktown area: just south of SR-72; at the Einstein
   Silver Mine.
    Arsenopyrite. Fluorite. Pyrite. Quartz, rock. Silver.
    Sphalerite. (2)

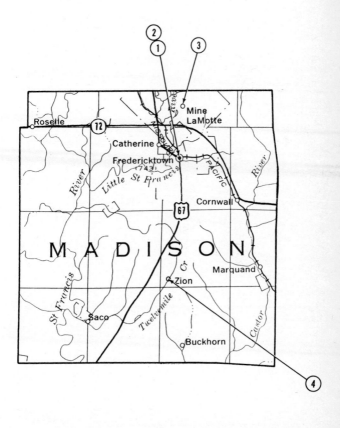

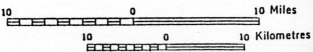

3. Mine La Motte area lead mines.
   Chalcopyrite. Galena. Lead. Pyrite. Sphalerite. (2)

4. Zion area: collect eastward to the Bollinger County border.
   Jasper. (1-4-13)

McDONALD COUNTY

1. Powell area: at Bee Bluff.
   Chert: high quality; pale blue; some pieces with drusy-
   lined cavities. (1-13-14-20)

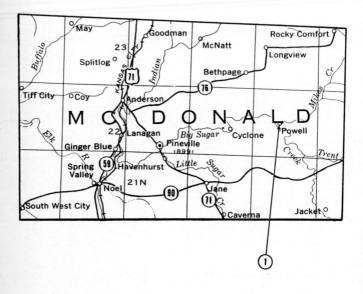

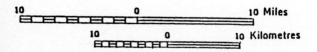

MILLER COUNTY

1. Etterville area mines.
   Barite: crystals. (2)

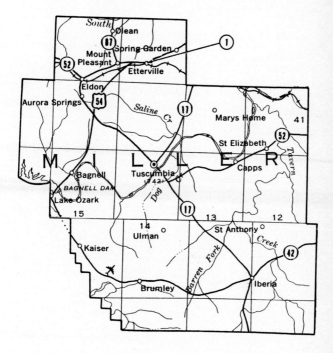

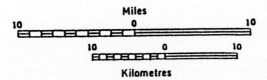

MONITEAU COUNTY

1. California area (and to some extent throughout remainder of county) gravel pits.

   Barite: good crystals. (10)

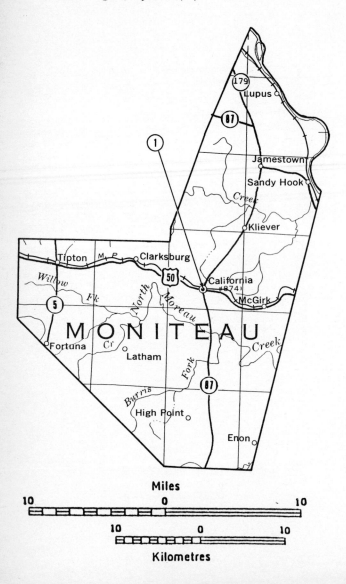

MORGAN COUNTY

1. Versailles area gravel pits.
   Barite: crystals. (10)

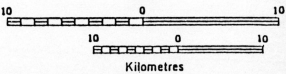

Miles

Kilometres

NEWTON COUNTY

1. Granby area mines.
   Calamine. Calcite. Cerussite. Chert. Dolomite: crystals. Galena. Marcasite. Pyrite. Quartz, rock. Sphalerite. (2)

2. Racine area quarries.
   Tripoli. (12)

3. Seneca area quarries.
   Tripoli. (12)

4. Wentworth area mines.
   Galena. Pyrite. Sphalerite. (2)

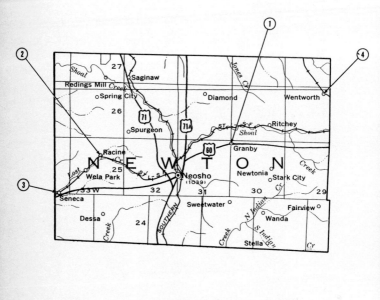

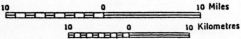

OZARK COUNTY

1. Gainesville area: 2 miles west on US-160; turn south on
   SR-5; follow to Timbered Knob.

    Chert: banded; white and yellow. (1-13)

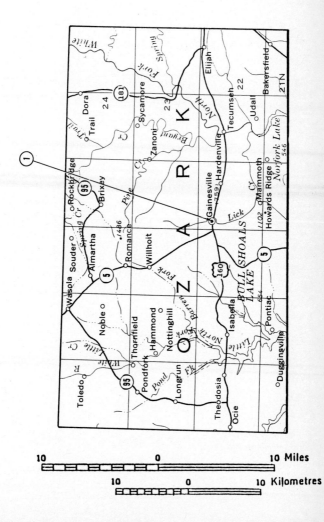

PHELPS COUNTY

1. Rolla area: 8 miles south on US-63 to dirt road at right; follow dirt road 2 miles west and bear right at fork; follow to Moselle Mine No. 10.

   Amethyst. Jasper. Quartz, rock. (2)

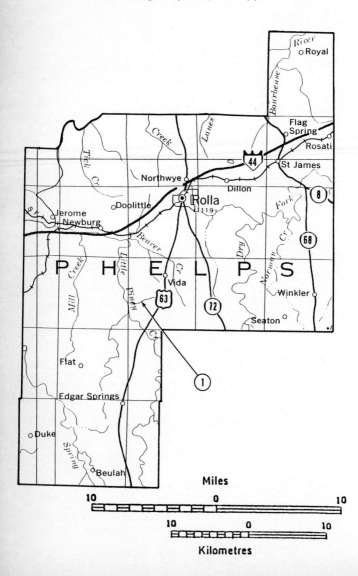

POLK COUNTY

1. County-wide in excavations, road cuts, RR cuts, and stream cuts along the west slope of the Ozark Mountains.

   Chert: colorful; some gemmy. (4-14-29)

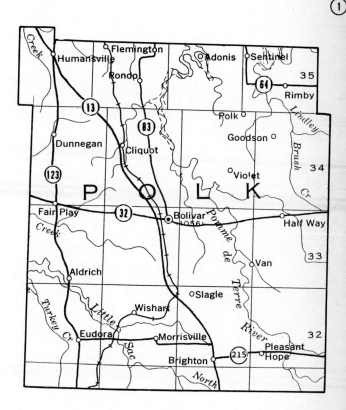

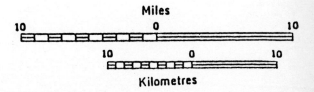

Miles

Kilometres

RALLS COUNTY

1. New London area: 1.6 miles north; in road cut at bridge over Salt River.

    Fossils: Paleozoic conodonts. (4)

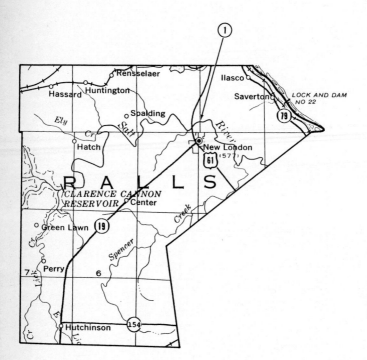

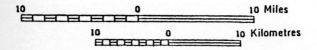

RIPLEY COUNTY

1. Doniphan area: collect along both sides of Current River upstream (north-northwest) to the Carter County border and downstream (south) to the Arkansas border.

    Chert: boulders; varied patterns and colors. (1-7-13-20)

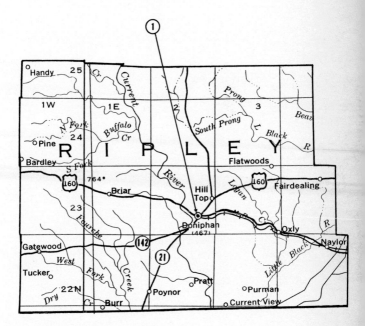

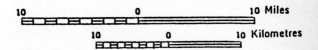

STE. GENEVIEVE COUNTY

1. Ozora area: 1.2 miles south; collect on both sides of Little Saline Creek upstream (southwest) to headwaters and downstream (northeast) to the Perry County border.

 Fossils, marine; chiefly invertebrates. (1-7-13-14-20)

ST. FRANCOIS COUNTY

1. Farmington area: 2 miles west on SR-32/CR-W to US-67; continue past US-67 on CR-W to first right; go one block north and turn left on paved road; go 2.2 miles west and bear right at fork; continue 0.7 mile to side road right (which may be blocked with chain); collect north and east from here.

 Chert (aka "Mozarkite"). (1-13)

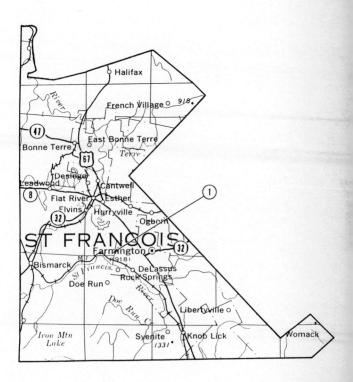

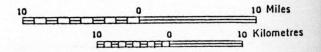

ST. LOUIS COUNTY

1. Creve Coeur area: at the Alton Brick Company.

   Fossils, marine: invertebrates; especially fine crinoid stems. (5-limestone)

2. Eureka area: in road, stream, and RR cuts.

   Fossils, marine: excellent trilobites; rolled and open. (4-5-shaly limestone -8-14-20)

3. Hazelwood area: 1.2 miles northeast on US-67 to St. Ferdinand Creek.

   Jasper. Quartz: geodes to 6″ diameter. (7-13-18-20)

4. St. Louis area quarries.
   Fossils, marine: fine echinoids (including the rare Melonechinus). (12)

SHANNON COUNTY

1. Eminence area: at the Jerktail Mine.
   Chalcocite. Malachite. (2)

2. Eminence area: at the Slater Mine.
   Chalcocite. Malachite. (2)

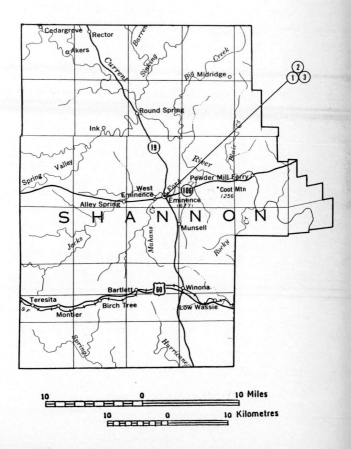

3. Eminence area: at the Tyrell Mine.
   Chalcocite. Malachite. (2)

STODDARD COUNTY

1. Bloomfield area clay pit.
   Fossils: flora; highly defined Eocene leaves; excellent
   quality. (25-35)

2. County-wide.
   Fossils: flora; quality Eocene period leaves. (25-35)

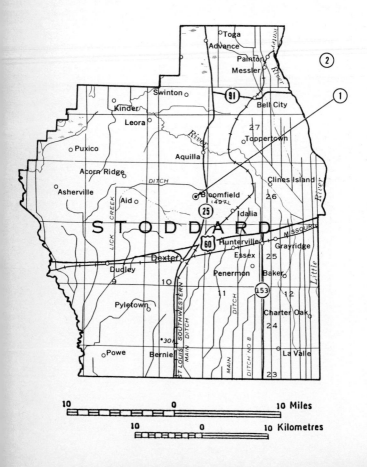

WASHINGTON COUNTY

1. Cadet area: at Old Mine.

   Agate. Barite: fine cabinet quality crystals; also cabinet quality massive, coated with glittering barite drusy and/or short small crystals. Chert: high quality: some with swirling patterns; varied pastel colors. Quartz, rock: as drusy. (2-38)

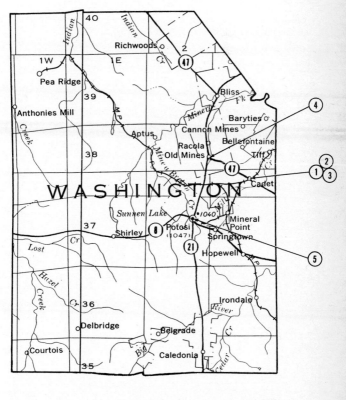

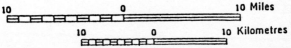

2. Cadet area: at Potosi Mine.

> Agate. Barite: fine cabinet quality crystals; also cabinet quality massive, coated with glittering barite drusy and/or short small crystals. Chert: high quality: some with swirling patterns; varied pastel colors. Quartz, rock: as drusy. (2-38)

3. Cadet area: at various mines and prospects.

> Agate. Barite: fine cabinet quality crystals; also cabinet quality massive, coated with glittering barite drusy and/or short small crystals. Chert: high quality: some with swirling patterns; varied pastel colors. Quartz, rock: as drusy. (2-21-38)

4. Old Mines area mines and prospects.

> Barite: fine cabinet quality crystals; also cabinet quality massive, coated with glittering barite drusy and /or short small crystals. Chert: high quality: some with swirling patterns; varied pastel colors. Quartz, rock: as drusy. (2-21-38)

5. Potosi area mines and prospects.

> Barite: good; some of cabinet quality. Chert: swirled patterns and varied pastel colors. Quartz, rock: usually as drusy. (2-21-38)

MONTANA

Statewide total of 133 locations in the following counties:

Beaverhead (6)	Gallatin (4)	Musselshell (1)
Big Horn (5)	Garfield (3)	Park (3)
Broadwater (8)	Granite (8)	Phillips (11)
Carbon (4)	Hill (1)	Powell (2)
Carter (1)	Jefferson (2)	Prairie (1)
Cascade (2)	Judith Basin (1)	Ravalli (1)
Chouteau (2)	Lewis and Clark	Richland (2)
Custer (3)	(14)	Rosebud (3)
Dawson (2)	Lincoln (5)	Silver Bow (12)
Deer Lodge (4)	Madison (9)	Stillwater (1)
Fallon (1)	Mineral (1)	Treasure (1)
Fergus (2)	Missoula (5)	Yellowstone (2)

STATEWIDE

Missouri River and tributaries, from North Dakota border to headwaters, as well as in other streams and some lakes.
Pearl. (7)

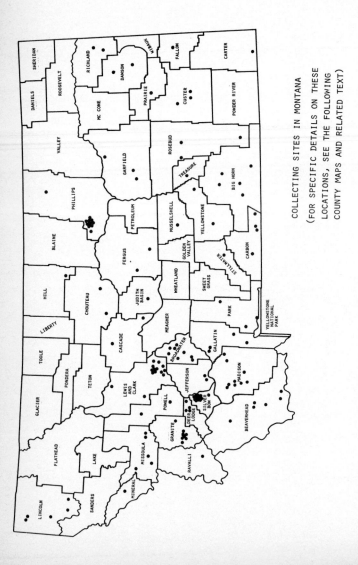

COLLECTING SITES IN MONTANA

(FOR SPECIFIC DETAILS ON THESE LOCATIONS, SEE THE FOLLOWING COUNTY MAPS AND RELATED TEXT)

BEAVERHEAD COUNTY

1. Dillon area: northwest; in Frying Pan Basin.
   Petrified wood: fine; silicified; opalized; some of gem quality; logs to 5′ diameter and 50′ long. (1-13)

2. Dillon area: 0.75 mile southeast on Sweetwater Road; at the Keystone Quarry.
   Talc (aka Soapstone). (12)

3. Elkhorn Springs area: 4 miles north; at Crystal Park.
   Amethyst. (1-13)

4. Lima area: downstream along the Red Rock River to Red Rock.
   Agate. Jasper. Petrified wood. (7-13-20)

5. Red Rock area: 5 miles north; on the Rebish ranch; in Sweetwater Creek.
   Corundum: asteriate; some gem quality. (1-7-13-20)

6. Snowline area: upstream along Sawmill Creek to the headwaters, just below (north of) the Continental Divide.
   Agate. Jasper. Petrified wood. (7-13-20)

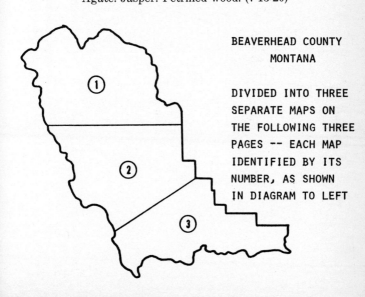

BEAVERHEAD COUNTY
MONTANA

DIVIDED INTO THREE
SEPARATE MAPS ON
THE FOLLOWING THREE
PAGES -- EACH MAP
IDENTIFIED BY ITS
NUMBER, AS SHOWN
IN DIAGRAM TO LEFT

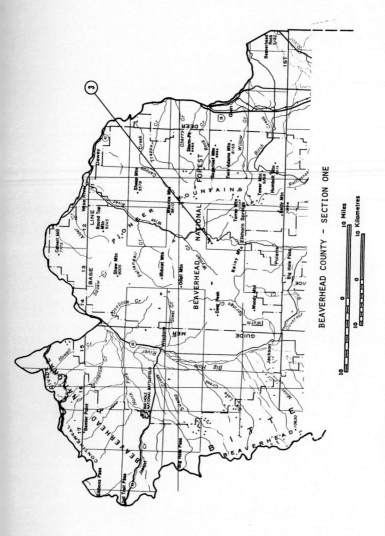

BEAVERHEAD COUNTY – SECTION ONE

BEAVERHEAD COUNTY - SECTION TWO

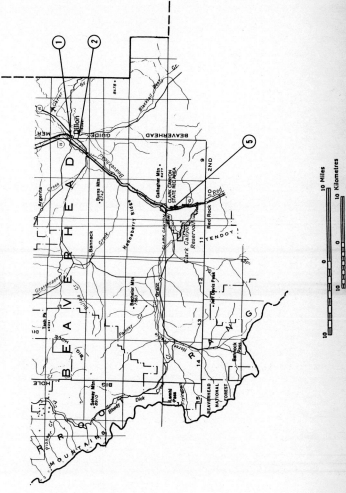

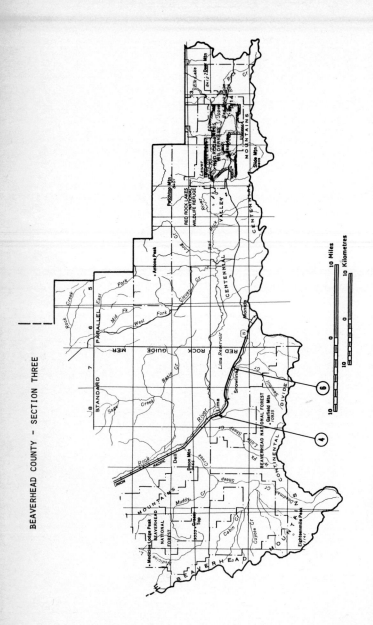

BEAVERHEAD COUNTY – SECTION THREE

BIG HORN COUNTY

1. Dunmore area: collect southward along the Burlington Northern RR tracks to Crow Agency; on east shore of Little Bighorn River.

    Agate. Jasper. Petrified wood. (20)

2. Fort Smith area: collect upstream along both sides of the Bighorn River to St. Xavier.

    Agate. Fossils: bones; some silicified. Jasper. Petrified wood. (8-14-20)

3. Hardin area: 1.75 miles north on SR-47 to bridge over Whitman Coulee; collect upstream (west and northwest) on both sides.

    Agate. Jasper. Petrified wood. (3-7-13-20)

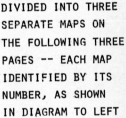

BIG HORN COUNTY
MONTANA

DIVIDED INTO THREE
SEPARATE MAPS ON
THE FOLLOWING THREE
PAGES -- EACH MAP
IDENTIFIED BY ITS
NUMBER, AS SHOWN
IN DIAGRAM TO LEFT

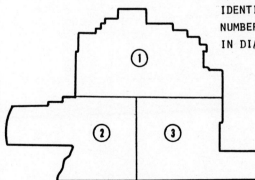

4. St. Xavier area: 2 miles south on the Fort Smith Road to side road left leading toward Lodge Grass; follow latter east and south another 8 miles to third bridge over Rotten Grass Creek; collect upstream on both sides of creek.

Agate. Jasper. Petrified wood. (1-3-7-13-20)

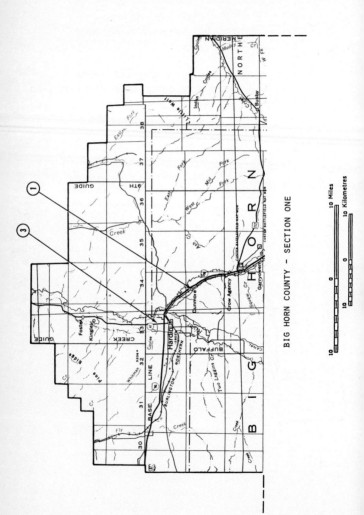

5. Wyola area: along rim and at base of steep canyons of Doghead Creek; most abundant near the TX Ranch at the mouth of Doghead Creek where it empties into the Bighorn River.

> Agate, banded: red, white; nodules to 9″ diameter. Chalcedony: geodes to 9″ diameter. (1-5-sandstone -13-19-20)

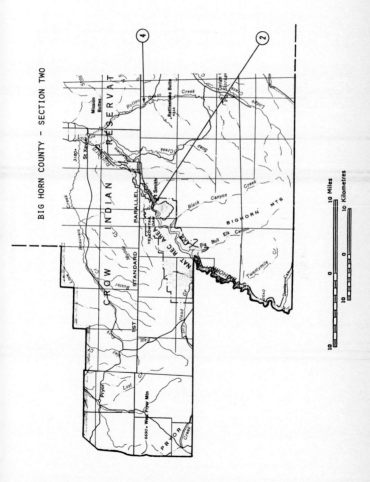

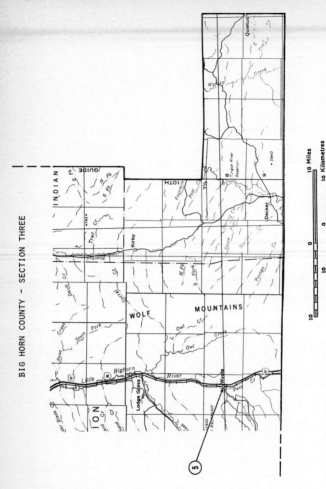

BROADWATER COUNTY

1. Radersburg area: at the Keating Mine.
   Gold. Silver. Tetradymite. (2)

2. Radersburg area mines.
   Gold. Silver. (2-19-andesite flows -22)

3. Townsend area: at the Hassel area mines.
   Gold. (2)

4. Townsend area: at the Park area mines.
   Gold. (2)

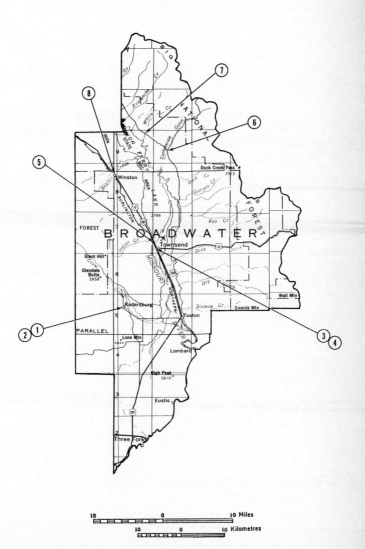

5. Townsend area: 1.2 miles northeast on US-12/287; then west to the Indian Creek area mines.

   Gold. (2)

6. Townsend area: 2 miles east on US-12 to left turn on CR; follow latter 14.25 miles north to side road at right that follows southeast side of Confederate Gulch; follow this northeast to area mines and placers.

   Gold. (17-21)

7. Townsend area: 18.25 miles north; 4 miles beyond (north of) Confederate Gulch to White Creek; collect at the White Creek area mines.

   Azurite. Chalcocite. Chalcopyrite. Chrysocolla. Copper. Cuprite. Malachite. (2)

8. Winston area: collect at the Beaver Creek area mines.

   Galena. Lead. Silver. Zinc. (2)

CARBON COUNTY

1. Hillsboro area: north to the Morrow/Frates diggings (CF); difficult road but not impassable for two-wheel-drive vehicles.

   Agate (Dryhead var.): nodules to 60 lb; geodes to 6″ diameter; fine quality. (1-13)

2. Red Lodge area: in prospect dumps on Hellroaring Plateau.

   Serpentine. (2)

3. Warren area: to Dryhead region and north from there on back road to foot of Pryor Gap; on rimrocks overlooking Bighorn Canyon; in adjacent grazing areas.

   Agate. (1-13)

4. Warren area: 5 miles north in Pryor Mountains.

   Agate. (1-13)

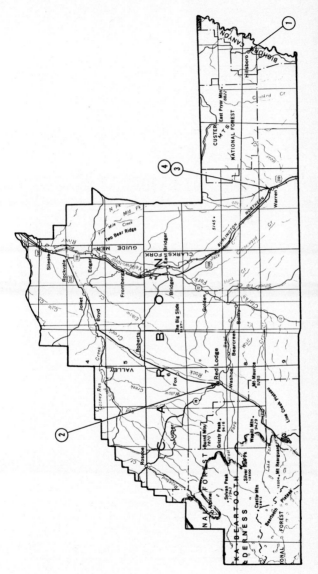

CARTER COUNTY

1. Alzada area deposits.
   Bentonite. (1-13-14)

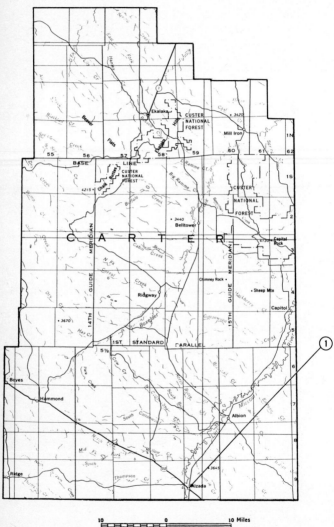

CASCADE COUNTY

1. Neihart area: at the Hartley Silver Mine.
   Silver. Sphalerite: green. (2)

2. Vaughn area: westward, along both sides of the Sun River
   and on both sides of US-89.
   Agate: black. Petrified wood: some with well-defined
   wood grain; some with worm borings. (1-7-13-20)

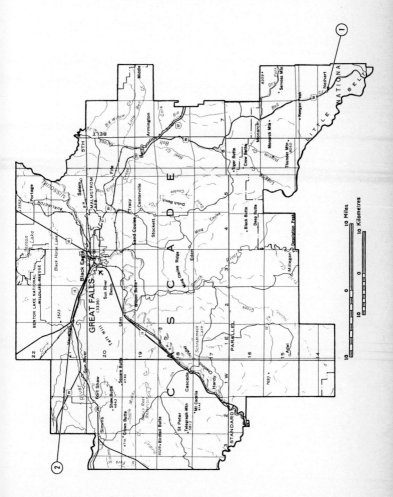

CHOUTEAU COUNTY

1. Highwood area: 0.6 mile west to bridge over Highwood Creek; upstream on both sides to base of Highwood Baldy.

    Agate: black, red. Petrified wood. (7-18-20)

2. Iliad area: 5 miles west and northwest on the Big Sandy Road to the second bridge over Chip Creek; upstream on both sides of creek to Eagleton area.

    Agate: black. Petrified wood. Sapphire. (3-7-13-18-20)

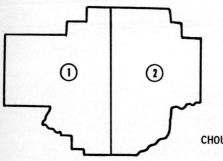

CHOUTEAU COUNTY
MONTANA

DIVIDED INTO TWO
SEPARATE MAPS ON
THE FOLLOWING TWO
PAGES -- EACH MAP
IDENTIFIED BY ITS
NUMBER, AS SHOWN
IN DIAGRAM TO LEFT

Montana 247

CHOUTEAU COUNTY
SECTION ONE

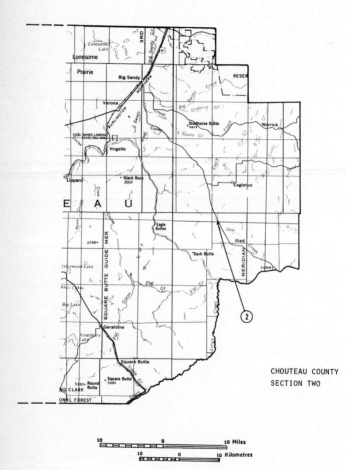

CHOUTEAU COUNTY
SECTION TWO

10   0   10 Miles
10   0   10 Kilometres

CUSTER COUNTY

1. County-wide; along both banks of the Yellowstone River from the Prairie County border upstream to the Rosebud County border; most particularly at the mouths of all tributaries, large or small.

    Agate, moss (aka Montana agate): very fine; nodules to 10″ diameter. (7-20)

2. Miles City area: 12 miles south on SR-59 to the Garland Road fork; bear right onto the Garland Road and go 0.75 mile to the Pumpkin Creek bridge; collect along the south bank, downstream to the creek's mouth at the Tongue River.

    Agate, moss (aka Montana agate): good quality; nodules to 8″ diameter. (7-20)

3. Mizpah area: 12 miles south-southeast on the Powderville Road to the Ash Creek bridge; collect downstream on both sides to the creek's mouth at the Powder River.

    Agate, moss (aka Montana agate): fine; nodules to 10″ diameter. (7-20)

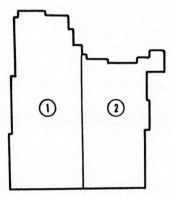

CUSTER COUNTY
MONTANA

DIVIDED INTO TWO
SEPARATE MAPS ON
THE FOLLOWING TWO
PAGES -- EACH MAP
IDENTIFIED BY ITS
NUMBER, AS SHOWN
IN DIAGRAM TO LEFT

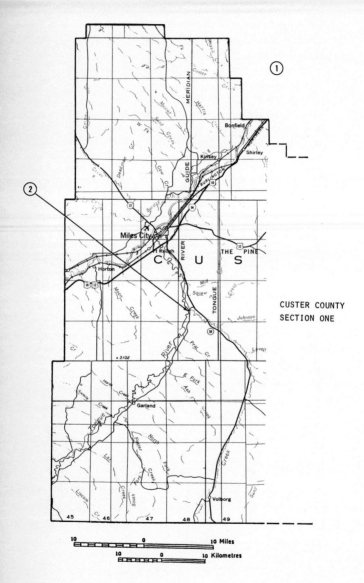

CUSTER COUNTY
SECTION ONE

CUSTER COUNTY - SECTION TWO

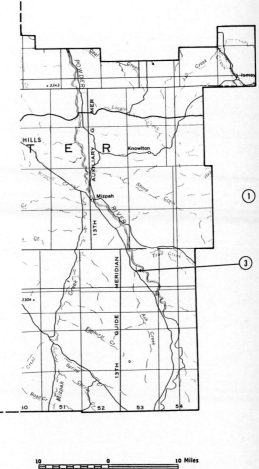

DAWSON COUNTY

1. Bloomfield area: 6.35 miles east on the Intake Road to the Thirteenmile Creek bridge; collect downstream on both sides of creek for 0.2 mile to mouth of tributary; especially at that confluence.

    Agate, moss (aka Montana agate). Jasper. Petrified wood. (7-20)

2. Glendive area: along both sides of the Yellowstone River (especially near mouths of tributaries) downstream (northeast) to the Richland County border and upstream (southwest) to the Prairie County border.

    Agate, moss (aka Montana agate). Jasper. Petrified wood. (7-20)

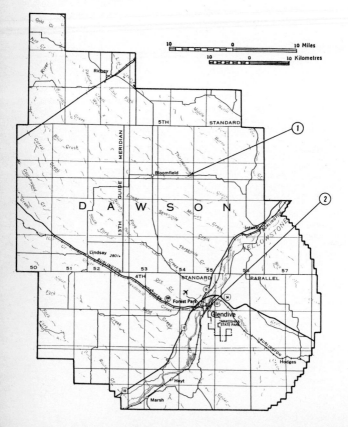

DEER LODGE COUNTY

1. Anaconda area: Blackfoot region; on Nelson Hill.

   Diamond: 6 stones found in 1883 and 1884, largest of which was a 0.22-ct flawed dodecahedron with triangular surface markings. (6-17)

2. Anaconda area: 2 miles east of the sign for French Gulch.

   Petrified wood. (1-13)

3. Anaconda area: 3 miles north; at Lost Creek Falls.

   Amazonite (aka Amazon stone). (16)

4. Galen area: 5 miles east of Champion district; at Dry Cottonwood Creek.

   Sapphire. (7)

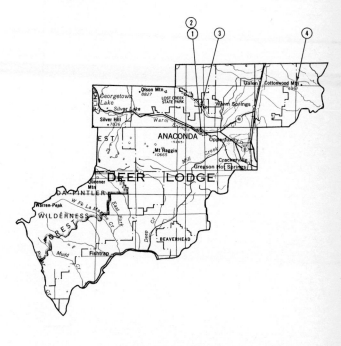

FALLON COUNTY

1. Baker area: in an arc running from 2 miles west on US-12
   to 2 miles south on SR-7.

   Fossils: excellent specimens of ammonites, ivory, and
   bones; all well silicified. Petrified wood. (1-3-13-14)

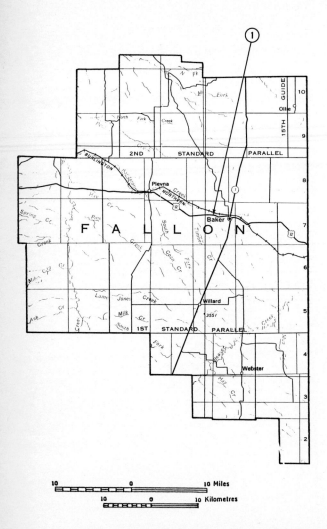

FERGUS COUNTY

1. Forestgrove area: 8 miles south and east to where small
   north tributary of the North Fork empties into the North
   Fork close to road; collect downstream along north bank
   of North Fork to where it is joined by South Fork to form
   Flatwillow Creek; continue collecting down north side of
   Flatwillow Creek (road continues generally alongside) all
   the way to US-87 (in Petroleum County), which is approximately 23 miles from point where collecting began.

   Agate. Fossils: ammonites, bones, some ivory; some
   well silicified. Jasper. Sapphire (unconfirmed). (1-7-13-
   20-29-39)

2. Kingston area: 4 miles west to Ross River bridge; collect
   downstream on both sides for 4 miles, to RR bridge in area
   of Ross Fork.

   Sapphire. (7-18-20)

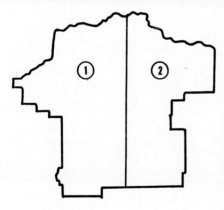

FERGUS COUNTY
MONTANA

DIVIDED INTO TWO
SEPARATE MAPS ON
THE FOLLOWING TWO
PAGES -- EACH MAP
IDENTIFIED BY ITS
NUMBER, AS SHOWN
IN DIAGRAM TO LEFT

FERGUS COUNTY
SECTION ONE

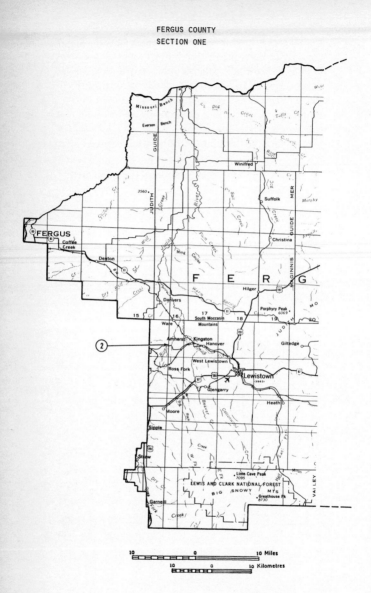

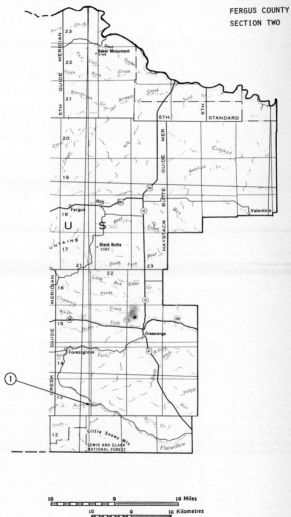

FERGUS COUNTY
SECTION TWO

10   0   10 Miles

10   0   10 Kilometres

GALLATIN COUNTY

1. Big Sky area.
   Corundum: fine gem quality; crystals to 14″ long.
   (1-7-13-20)

2. Bozeman area: 15 miles due south; on Mount Blackmore,
   close to summit.
   Chalcedony. Hyalite: fine. Opal, common: fine; translucent orange. (1-13)

3. Logan area.
   Calcite (Onyx var., aka Montana onyx); well patterned
   and finely banded. (1-13)

4. Logan area: in bluffs where Elk Creek empties into the
   Madison River.
   Petrified wood: very fine. (5-sandstone -8-13-33)

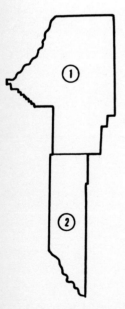

GALLATIN COUNTY
MONTANA

DIVIDED INTO TWO
SEPARATE MAPS ON
THE FOLLOWING TWO
PAGES -- EACH MAP
IDENTIFIED BY ITS
NUMBER, AS SHOWN
IN DIAGRAM TO LEFT

GALLATIN COUNTY
SECTION ONE

10   0   10 Miles

10   0   10 Kilometres

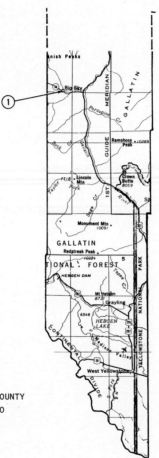

GALLATIN COUNTY
SECTION TWO

GARFIELD COUNTY

1. Edwards area: upstream and down in Big Dry Creek.
   Agate. Jasper. Opal (unconfirmed). Petrified wood.
   (1-7-13-20)

2. Jordan area: 22.6 miles north on CR; then 0.75 mile east
   toward arm of Fort Peck Reservoir.
   Fossils: fish; reptiles; bones; figs. Petrified wood.
   (1-13-14-29-39)

3. Sand Springs area: downstream along South Fork of
   Lodgepole Creek.
   Agate. Jasper. Opal (unconfirmed). Petrified wood.
   Sapphire (unconfirmed). (1-7-13-20)

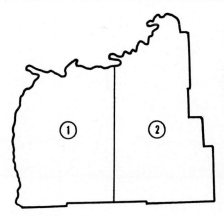

GARFIELD COUNTY
MONTANA

DIVIDED INTO TWO
SEPARATE MAPS ON
THE FOLLOWING TWO
PAGES -- EACH MAP
IDENTIFIED BY ITS
NUMBER, AS SHOWN
IN DIAGRAM TO LEFT

GARFIELD COUNTY
SECTION ONE

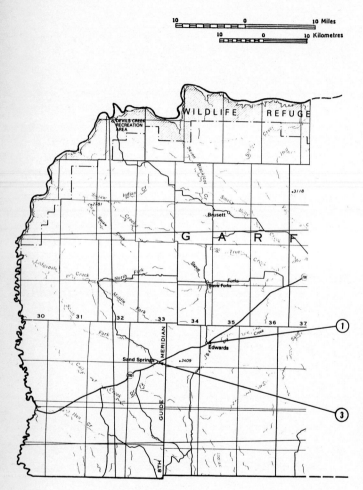

GARFIELD COUNTY
SECTION TWO

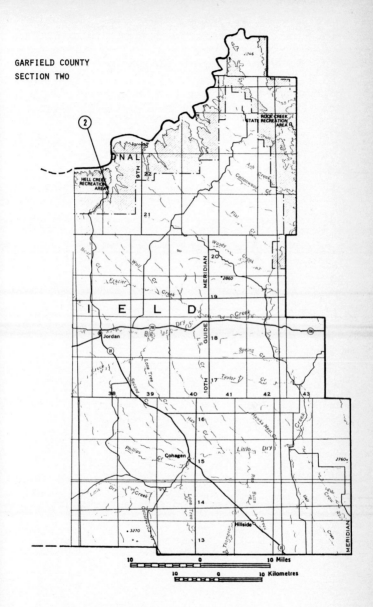

ROCK CREEK
STATE RECREATION
AREA

HELL CREEK
RECREATION
AREA

DNAL

9TH

22

21

Bear Cr.

Woll

Cr.

Glacier

Cr.

Creek

Woody

Creek

20

*2860

19

I E L D

Creek

Dry

Big

Creek

Jordan

18

GUIDE

MERIDIAN

Spring

Cr.

Ash

Creek

Cottonwood

Cr.

Flat

Cr.

Lone Tree

Cr.

Second

Cr.

Creek

38

39

40

Taylor

10TH

17

Cr.

41

Hawks Nest Cr.

Cr.

42

Creek

43

16

Hay

Cr.

Little

Dry

Phillips

Cr.

Cohagen

15

Red

Cr.

2760*

Little

Dry

Creek

Lone

Tree

Cr.

Cottonwood

Cr.

14

*3270

13

Thompson Cr.

Hillside

Butte

Creek

Crow

Creek

Uti.

Crow River

MERIDIAN

10   0   10 Miles

10   0   10 Kilometres

GRANITE COUNTY

1. Philipsburg area mines.

   Rhodochrosite: massive; pale pink. (2)

2. Philipsburg area: 7 miles south on US-10A to right turn on SR-38; 11 miles west on SR-38 to the Cornish Sapphire Mine (CF).

   Ruby: gem quality crystals. Sapphire: gem quality crystals. (2)

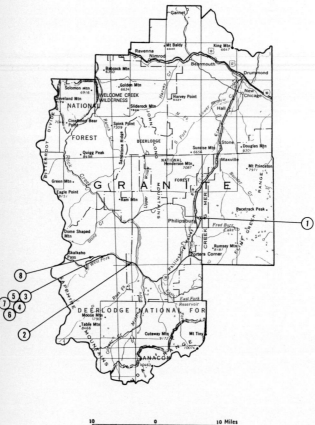

3. Philipsburg area: 7 miles south on US-10A to right turn on SR-38; continue on that for 15 miles to bridge across West Fork; follow signs to Anaconda Gulch (CF).

    Sapphire: gem quality. (1-5-7-9-14-29)

4. Philipsburg area: 7 miles south on US-10A to right turn on SR-38; continue on that for 15 miles to bridge across West Fork; follow signs to Basin Gulch (CF).

    Sapphire: gem quality. (1-5-7-9-14-29)

5. Philipsburg area: 7 miles south on US-10A to right turn on SR-38; continue on that for 15 miles to bridge across West Fork; follow signs to Coe Gulch.

    Sapphire: gem quality. (1-5-7-9-14-29)

6. Philipsburg area: 7 miles south on US-10A to right turn on SR-38; continue on that for 15 miles to bridge across West Fork; follow signs to Malay Gulch (CF).

    Sapphire: gem quality. (1-5-7-9-14-29)

7. Philipsburg area: 7 miles south on US-10A to right turn on SR-38; continue on that for 15 miles to bridge across West Fork; follow signs to Sapphire Gulch (CF).

    Sapphire: gem quality. (1-5-7-9-14-29)

8. Philipsburg area: 7 miles south on US-10A to right turn on SR-38; 17 miles west on SR-38 to the Chausee Sapphire Mine on Rock Creek.

    Sapphire. (2)

HILL COUNTY

1. Rocky Boy area: in the Bearpaw Mountains.

    Columbite. (29-carbonate rocks)

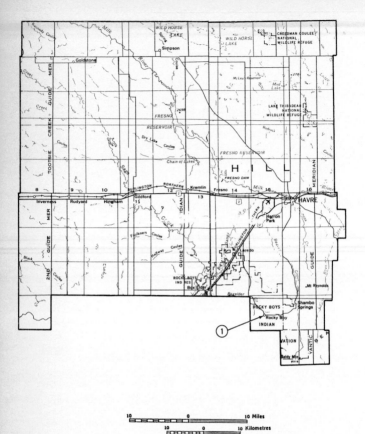

JEFFERSON COUNTY

1. Whitehall area: west on US-10 to where Toll Mountain
   Road (a dirt road) turns north; follow it to Toll Mountain
   Picnic Grounds; 2 miles northeast of the picnic grounds on
   a forested ridge; at the Pohndorf Amethyst Mine.

   Amethyst: fine. Feldspar: crystals to 14″. Quartz,
   rock: crystals to 8″ diameter and 3′ long; some doubly
   terminated; some with amethystine terminations.
   (2-16)

2. Whitehall area: 18 miles northwest; in Hay Canyon; at the Gem Queen claim.

   Quartz, smoky. Titanite (aka Sphene): quality; translucent brown; crystals to 1″ thick. Tourmaline: black. (1-2)

JUDITH BASIN COUNTY

1. Utica area: 10.4 miles southwest on gravel road to foot of
   the Little Belt Mountains; in Judith Basin; collect in 5-mile
   area of Yogo Gulch. (CF in one area.)

   > Aegirite-diopside. Analcite. Apatite. Biotite. Calcite.
   > Dolomite. Feldspar. Hematite. Magnetite. Sapphire:
   > very fine gem quality; cornflower blue. Zircon. (7-16)

LEWIS AND CLARK COUNTY

1. Augusta area: upstream (southwest) on both sides of Elk Creek to mouth of Smith Creek.

    Agate. Gold. Jasper. Sapphire (unconfirmed). Petrified wood. (1-3-7-13-20)

2. Canyon Ferry area: just south to Canyon Ferry Dam area; in Missouri River 1 mile upstream from Magpie Gulch.

    Sapphire: green. (7-18-20)

3. East Helena area: from west edge of town, go north on SR-433 to right turn at Canyon Ferry Road; east on Canyon Ferry Road to York Road; 1 mile north on York Road to side road left; follow this west to the Guffey Sapphire Mine (CF).

    Sapphire. (2)

4. Helena area: at Scratch Gravel.

    Vesuvianite: quality crystals to 1.5″ thick. (2-5-in powder/blue calcite)

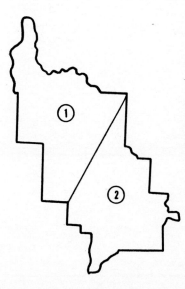

LEWIS AND CLARK COUNTY
MONTANA

DIVIDED INTO TWO
SEPARATE MAPS ON
THE FOLLOWING TWO
PAGES -- EACH MAP
IDENTIFIED BY ITS
NUMBER, AS SHOWN
IN DIAGRAM TO LEFT

5. Helena area: north on Montana Avenue to right turn at Northgate Plaza on SR-280 (aka Custer Avenue East); 10.3 miles east to right turn at Hart Lane; very quickly turn right again at old white schoolhouse and then make an immediate left turn, crossing over cattle guard; follow this to its end at the Castle Sapphire Mine (CF).

Sapphire. (2)

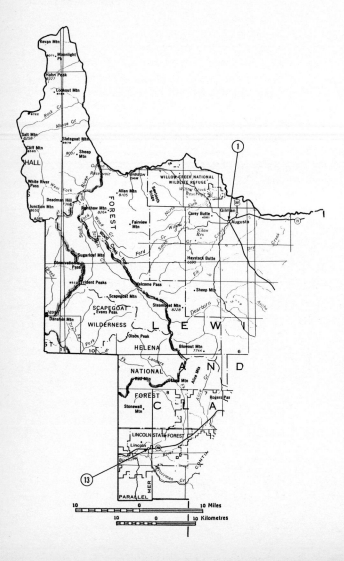

6. Helena area: 3 miles southeast; at Three-Mile Gulch.
   Quartz, smoky. (1-13)

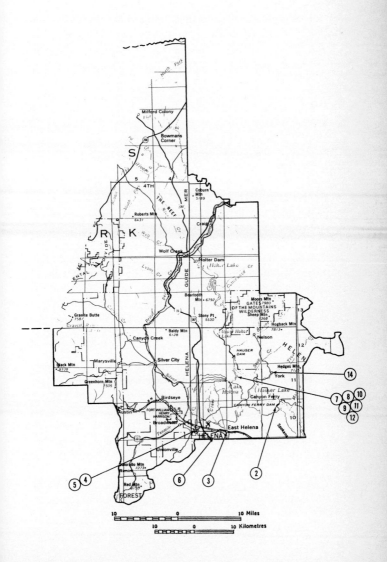

7. Helena area: 10.5 miles northeast; upstream from Eldorado Bar in the Missouri River; on Emerald Bar.

> Cassiterite (aka Stream tin). Chalcedony. Garnet. Gold. Kyanite. Limonite. Pyrite. Sapphire. Topaz. (7-18)

8. Helena area: 11 miles northeast; upstream from Eldorado Bar in the Missouri River; in upper end of Magpie Gulch.

> Cassiterite (aka Stream tin). Chalcedony. Garnet. Gold. Kyanite. Limonite. Pyrite. Sapphire. Topaz. (7-18)

9. Helena area: 12 miles northeast; on Eldorado Bar in the Missouri River.

> Cassiterite (aka Stream tin). Chalcedony. Garnet. Gold. Kyanite. Limonite. Pyrite. Sapphire. Topaz. (7-18)

10. Helena area: 12 miles northeast; vicinity of Eldorado Bar in the Missouri River; on Spokane Bar.

> Gold. Sapphire: pale blue. (7-18)

11. Helena area: 12 miles northeast; vicinity of Eldorado Bar in the Missouri River; in and near the mouth of Prickly Pear Creek.

> Gold. Sapphire: pale blue. (7-18)

12. Helena area: 14 miles northeast; on the American Bar, downstream from Eldorado Bar.

> Cassiterite (aka Stream tin). Chalcedony. Garnet. Gold. Kyanite. Limonite. Pyrite. Sapphire. Topaz. (7-18)

13. Lincoln area streams.

> Agate. Garnet. Gold. Petrified wood. Sapphire. (1-3-7-13-20)

14. York area: go to gas station where road goes north; 5 miles north on that road to a dirt road left; 4 miles southwest on dirt road to El Dorado Sapphire Diggings (CF).

> Garnet. Sapphire. (1-3-7-13-18)

LINCOLN COUNTY

1. Libby area: 4 miles southwest; at the Rainy Creek mines.
    Aegirite. Cuprite. Gold. Hematite. Lead. Quartz, milky. Silver. Tremolite. Vermiculite. (2)

2. Libby area: 20 miles south; in mines.
    Gold. Lead. Silver. (2-19-quartz -21)

3. Libby area: 35 miles southeast; in mines.
    Argillite. Barite. Gold. Lead. Silver. (2-21)

4. Yaak area: to the Sylvanite area mines and prospects along the Yaak River.
    Gold. Pyrite. Sphalerite. (2-21)

5. Yaak area mines and prospects along the Yaak River.
    Gold. Pyrite. Sphalerite. (2-21)

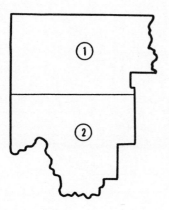

LINCOLN COUNTY
MONTANA

DIVIDED INTO TWO
SEPARATE MAPS ON
THE FOLLOWING TWO
PAGES -- EACH MAP
IDENTIFIED BY ITS
NUMBER, AS SHOWN
ON DIAGRAM TO LEFT

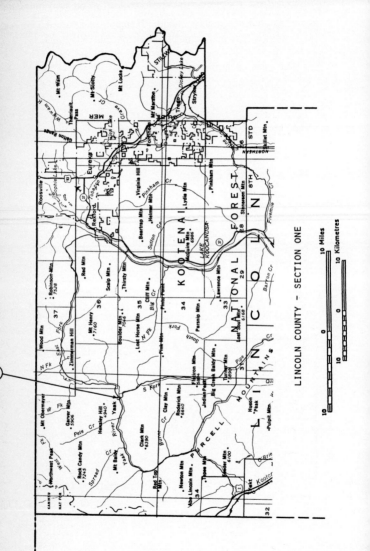

LINCOLN COUNTY – SECTION ONE

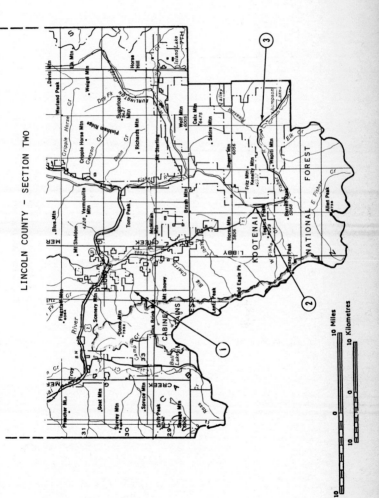

LINCOLN COUNTY – SECTION TWO

MADISON COUNTY

1. Greenhorn Gulch area (more specific location unknown).

    Diamond: 1 "pea-sized" gem quality stone found during gold-panning operations; other specifics unknown. (17)

2. Red Bluff area: near Finnegan Ridge and Elk Mountain; at Cow Camp.

    Chert: fine. Jasper: green, red; good quality. (1-13)

3. Red Bluff area: 6 miles east and northeast on SR-84 to Pole Creek bridge; collect on both sides downstream 0.25 mile to the creek's mouth at the Madison River.

    Garnet. Gold. Ruby (Corundum): one chunk found weighing 588.5 ct. Sapphire: gem quality; blue. (7-20)

4. Ruby Dam area: collect downstream on both sides of river all the way to the Alder area.

    Garnet, almandine: abundant; good quality. (1-7-13-20)

5. Ruby Dam area: collect upstream on both sides of reservoir.

    Garnet, almandine: abundant; good quality. (7-20)

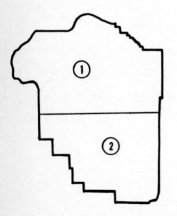

MADISON COUNTY
MONTANA

DIVIDED INTO TWO
SEPARATE MAPS ON
THE FOLLOWING TWO
PAGES -- EACH MAP
IDENTIFIED BY ITS
NUMBER, AS SHOWN
IN DIAGRAM TO LEFT

6. Ruby River Dam area: 1.5 miles upstream (south) on Ruby River Road to first large tributary entering reservoir from east; collect along shores of that stream, upstream and down to the reservoir.

Garnet, almandine: good quality. (7-20)

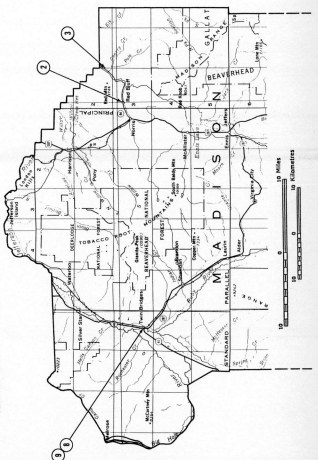

MADISON COUNTY
SECTION ONE

7. Ruby Dam area: south 6.3 miles to fork; bear right (heading southwest) and collect on both sides of the road all the way (23 miles) to the Beaverhead County border.

   Rhyolite (aka Wonderstone): concentrically banded. (1-4-13-37)

8. Twin Bridges area: west; in the foothills flanking the Jefferson River.

   Petrified wood: fine. (1-3-7-13-20)

9. Twin Bridges area: west to Rochester River area: at Crystal Butte.

   Quartz, rock. (1-13)

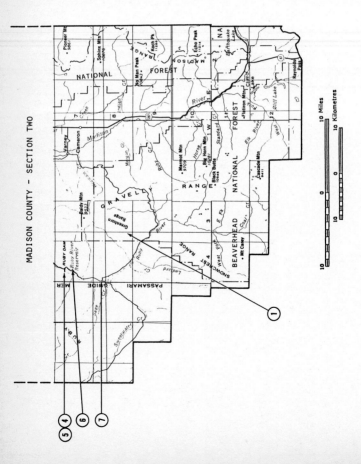

MINERAL COUNTY

1. Superior area: at the Iron Mountain Mine.
   Boulangerite. (2)

MISSOULA COUNTY

1. Greenough area mines and placers on Elk Creek.
   Gold. Pyrite. Quartz, rock. (2-19-quartz -21)

2. Greenough area: 2 miles south of Elk Creek; in deposits.
   Barite. (1-13)

3. Lolo area: along Lolo Creek.
   Quartz, rock. Quartz, smoky. (1-13)

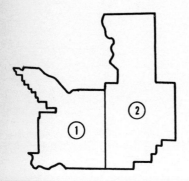

MISSOULA COUNTY
MONTANA

DIVIDED INTO TWO
SEPARATE MAPS
-- EACH MAP
IDENTIFIED BY ITS
NUMBER, AS SHOWN
IN DIAGRAM TO LEFT

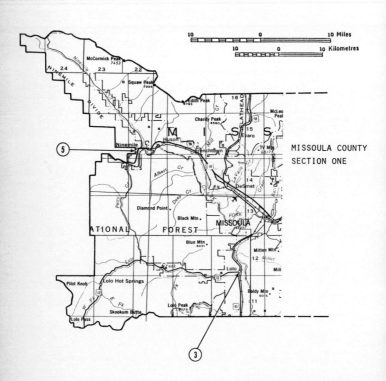

MISSOULA COUNTY
SECTION ONE

4. Mineral Peak area mines.
   Arsenopyrite. Silver. (2)

5. Ninemile area.
   Gold. (3-glacial moraines -17-20)

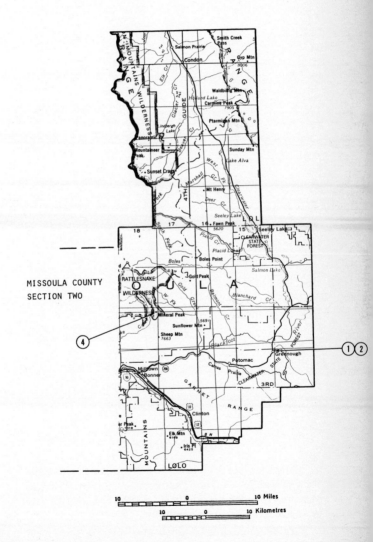

MUSSELSHELL COUNTY

1. Roundup area: 2 miles north on US-87; just east of highway.

    Fossils, marine: some agatized. (1-13)

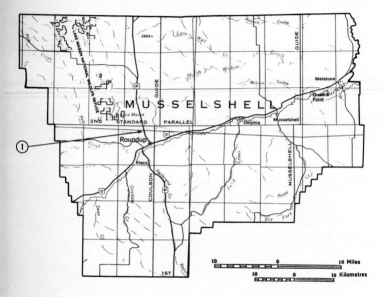

PARK COUNTY

1. Gardiner area: north on US-89 to left turn leading to Gallatin National Forest Campground; follow this road 10 miles
   to campground; park and walk northward up foot trail
   until above bluffs; then northwest.

    Agate: good quality. Opal, wood. (1-13)

2. Miner area.

   Petrified wood. (1-13)

3. Springdale area: 3 miles southwest on I-90 (US-10/191); along shoreline of Yellowstone River.

   Petrified wood: good quality. (3-7-18-20)

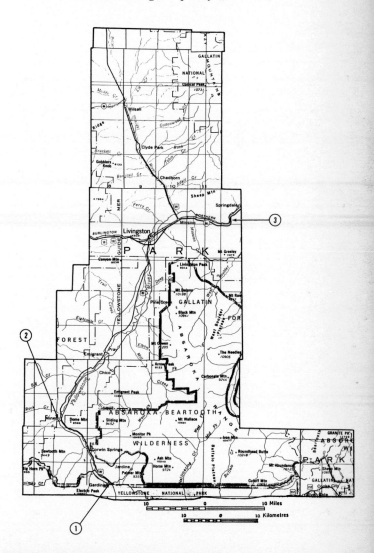

PHILLIPS COUNTY

1. Landusky area mines.
   Gold. Malachite. Pyrolusite. (2)

2. Whitewater area: collect upstream and down along both sides of Whitewater Creek.
   Agate. Petrified wood. (1-7-13-20)

3. Zortman area: at the Alabama Mine.
   Gold. (2)

4. Zortman area: at the Ella C Mine.
   Gold. (2)

5. Zortman area: at the Fergus Mine.
   Gold. (2)

6. Zortman area: at the Goldbug Mine.
   Gold. (2)

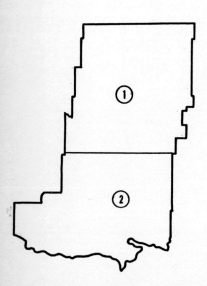

PHILLIPS COUNTY
MONTANA

DIVIDED INTO TWO
SEPARATE MAPS ON
THE FOLLOWING TWO
PAGES -- EACH MAP
IDENTIFIED BY ITS
NUMBER, AS SHOWN
IN DIAGRAM TO LEFT

7. Zortman area: at the Hawkeye Mine.
   Gold. (2)

8. Zortman area: at the Independent Mine.
   Gold (2)

9. Zortman area: at the Mint Mine.
   Gold (2)

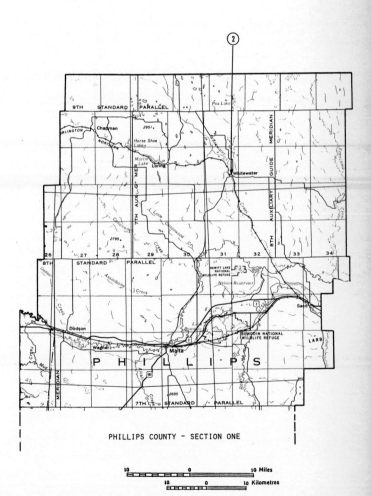

PHILLIPS COUNTY – SECTION ONE

10. Zortman area: at the Pole Gulch Mine.
   Gold (2)

11. Zortman area: in Alder Gulch; at the August Mine.
   Gold (2)

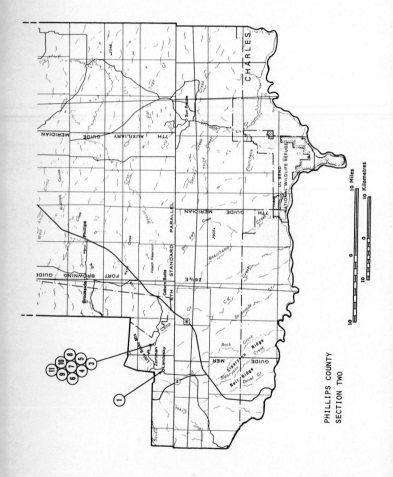

PHILLIPS COUNTY
SECTION TWO

POWELL COUNTY

1. Deer Lodge area: 4.4 miles east-southeast; in Dry Cotton-wood Creek.

    Sapphire: quality; blue. (3-7-8-18-20)

2. Ovando area: 6 miles east and southeast on SR-200 to east shore of Kleinschmidt Lake; collect on all shorelines, but especially on east side of lake.

    Agate. Jasper. Sapphire (occasional). (20)

POWELL COUNTY
MONTANA

DIVIDED INTO TWO
SEPARATE MAPS ON
THE FOLLOWING TWO
PAGES -- EACH MAP
IDENTIFIED BY ITS
NUMBER, AS SHOWN
IN DIAGRAM TO LEFT

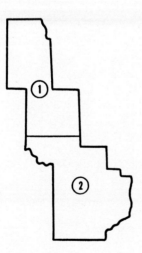

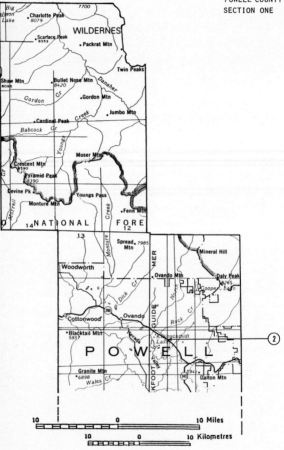

Elevation Mtn
7047
STANDARD
BLA[
Helmville
Douglas
Hoodoo Mtn
Nevada
Creek
Cottonwood Cr
Devil Mtn
7438
Old Baldy Mtn
7511
Saddle Mtn
6814
Hoover Cr
Brock Cr
White Rocks Mtn
6615
GARNET RANGE
GUIDE
Avon Valley
141
Goldcreek
Gold Cr
Little
Blackfoot
Avon
12
Snowshoe Cr
Blossburg
Dry Cr
Elliston
River
Lone Tree Hill
6267
Garrison
10
Rocky Ridge
5972
2ND STANDARD PARALLEL
BLACKFOOT
HELENA
Negro Mtn
Kohrs
Spotted Dog Cr
Pikes Peak
9035
Rock Cr
Fork
MERIDIAN
BN
Deer Lodge
4521
Clark
Peterson Cr
Bison Mtn
8028
Cliff Mtn
8301
Dempsey Cr
Racetrack
Racetrack
Dempsey

POWELL COUNTY - SECTION TWO

10 0 10 Miles
10 0 10 Kilometres

PRAIRIE COUNTY

1. Terry area: along both shores of the Yellowstone River, especially near mouths of tributaries; upstream to the Custer County border and downstream to the Dawson County border.

   Agate, moss (aka Montana agate): nodules to 10″ diameter. (7-18-20)

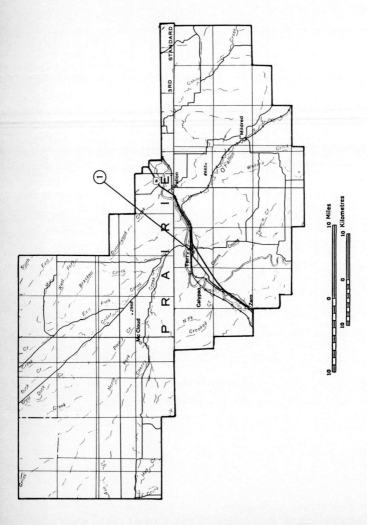

RAVALLI COUNTY

1. Sula area: 2 miles north on dirt road.
   Beryl. (16)

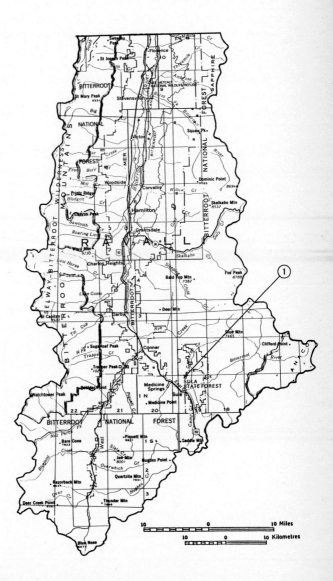

RICHLAND COUNTY

1. Crane area: 1 mile east; in Yellowstone River gravel bars.
   Agate, moss (aka Montana agate). (3-18-20)

2. Sidney area: along the shores of the Yellowstone River,
   especially at the mouths of tributaries; upstream to the
   Dawson County border and downstream to the North
   Dakota border

   Agate, moss (aka Montana agate): nodules to 10″ di-
   ameter. (7-20)

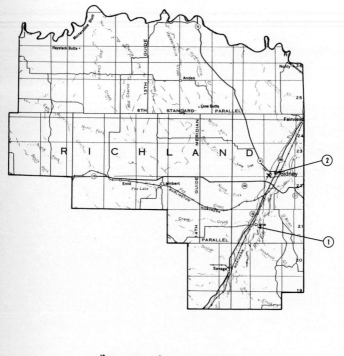

ROSEBUD COUNTY

1. Birney area: collect upstream and down along west shore of Tongue River, especially at mouths of tributaries.

    Agate, moss (aka Montana agate). Jasper. Petrified wood. (20)

2. Forsyth area: along the shores of the Yellowstone River, especially in the area of the mouths of tributaries; upstream to the Treasure County border and downstream to the Custer County border.

    Agate, moss (aka Montana agate): nodules to 10″ diameter. (7-20)

3. Vananda area: collect downstream along both sides of Horse Creek to its mouth at the Yellowstone River.

    Agate, moss (aka Montana agate): nodules to 10″ diameter. (20)

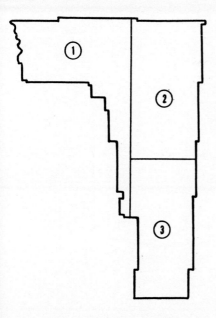

ROSEBUD COUNTY
MONTANA

DIVIDED INTO THREE
SEPARATE MAPS ON
THE FOLLOWING THREE
PAGES -- EACH MAP
IDENTIFIED BY ITS
NUMBER, AS SHOWN
IN DIAGRAM TO LEFT

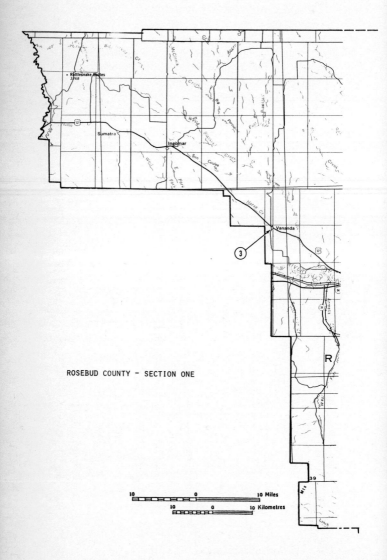

ROSEBUD COUNTY - SECTION ONE

10 [====] 0 [====] 10 Miles
10 [====] 0 [====] 10 Kilometres

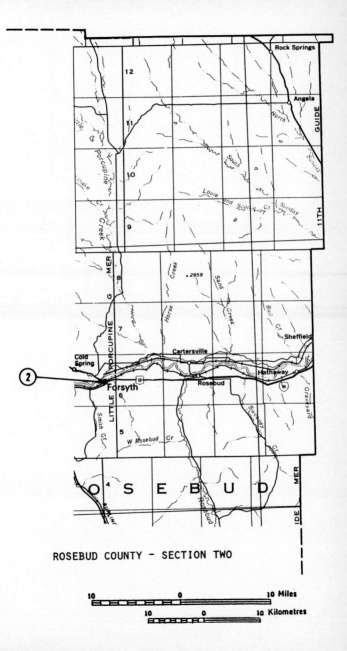

ROSEBUD COUNTY - SECTION TWO

10 ▭ 0 10 Miles

10 ▭ 0 10 Kilometres

ROSEBUD COUNTY - SECTION THREE

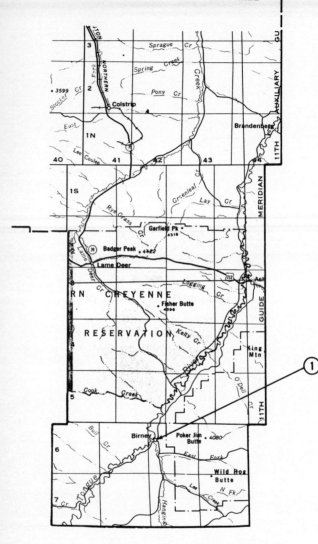

10 Miles

10 Kilometres

SILVER BOW COUNTY

1. Butte area: at the Alice Mine (inactive.)
     Rhodochrosite. Rhodonite. Silver. (2-19)

2. Butte area: at the East Colusa Mine.
     Covellite: opaque purple; tabular crystals in clusters.
     (2)

3. Butte area: at the Emma Mine.
     Rhodochrosite. (2)

4. Butte area: at the Kelley Mine.
     Melanterite. (2)

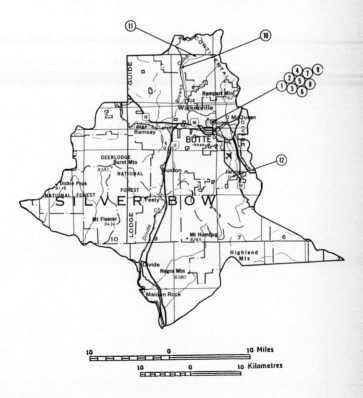

5. Butte area: at the Leonard Mine.

   Covellite: opaque purplish pink; tabular crystals in clusters. Pyrite: brassy; striated; cubic crystals in clusters. (2)

6. Butte area: in the Lexington Mine dumps.

   Rhodonite. (2)

7. Butte area: east to Little Pipestone Creek.

   Amethyst. (7-18-20)

8. Butte area: east to the west fork of Rader Creek.

   Amethyst. (7-18-20)

9. Butte area: south; in the Gravelley Range.

   Rhyolite: silicified. (1-13)

10. Butte area: 10 miles north-northwest; in Browns Gulch.

    Sapphire. (1-3-7-13-23)

11. Butte area: 12.5 miles north-northwest; in upper 4 miles of the south fork of Dry Cottonwood Creek.

    Sapphire. (7-13)

12. Janney area: slightly east of RR cut.

    Quartz, smoky. (4)

STILLWATER COUNTY

1. Columbus area: in vicinity of bridge over the Yellowstone River and on nearby islands.

   Jasper, moss. Petrified wood. (3-7-18-20)

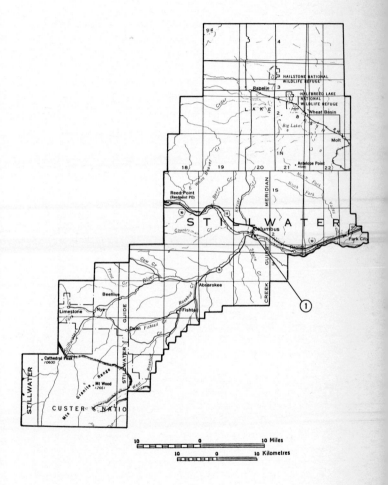

TREASURE COUNTY

1. Myers area: along shorelines of both banks of the Yellowstone River, especially at the mouths of tributaries; upstream to the Yellowstone County border and downstream to the Rosebud County border.

   Agate, moss (aka Montana agate): nodules to 10″ diameter. (7-20)

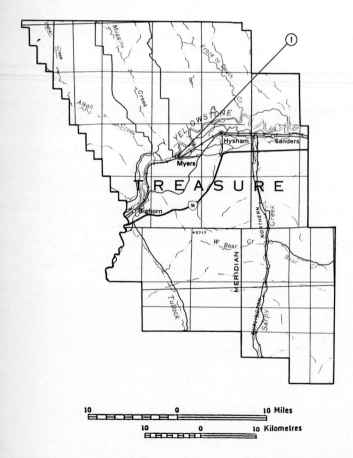

YELLOWSTONE COUNTY

1. Billings area: along both sides of the Yellowstone River, especially at the mouths of tributaries; downstream to the Treasure County border and upstream to the western border of Yellowstone County, where the river upstream forms the border of both Stillwater and Carbon counties.

   Agate, moss (aka Montana agate): nodules to 10″ diameter. (7-20)

2. Custer area: 2.3 miles west to the mouth of Buffalo Creek; collect along both sides of Buffalo Creek upstream (northwest) to the mouth of East Buffalo Creek and then for 5 miles up the latter creek.

   Agate, moss (aka Montana agate): nodules to 10″ diameter. (1-3-7-13-20)

YELLOWSTONE COUNTY
MONTANA

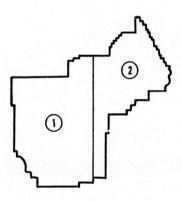

DIVIDED INTO TWO
SEPARATE MAPS ON
THE FOLLOWING TWO
PAGES -- EACH MAP
IDENTIFIED BY ITS
NUMBER, AS SHOWN
IN DIAGRAM TO LEFT

YELLOWSTONE COUNTY
SECTION ONE

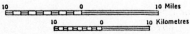

YELLOWSTONE COUNTY
SECTION TWO

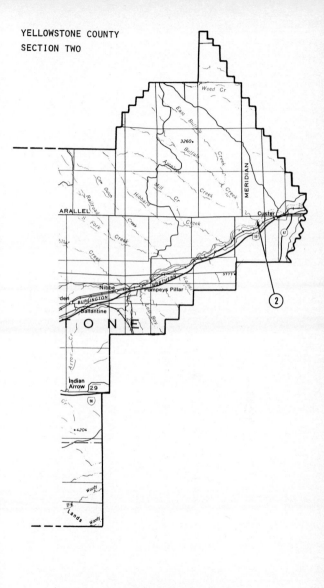

10   0   10 Miles
10   0   10 Kilometres

NEBRASKA

Statewide total of 143 locations in the following counties:

Adams (2)	Dodge (1)	Knox (1)
Blaine (1)	Douglas (2)	Lancaster (1)
Boyd (1)	Franklin (4)	Lincoln (3)
Brown (1)	Frontier (1)	Madison (1)
Buffalo (2)	Furnas (4)	Morrill (3)
Butler (1)	Gage (3)	Nance (2)
Cass (7)	Garden (1)	Nemaha (4)
Chase (1)	Hall (1)	Nuckolls (3)
Cherry (8)	Harlan (4)	Otoe (6)
Cheyenne (3)	Hayes (1)	Pawnee (2)
Clay (1)	Holt (1)	Phelps (1)
Custer (1)	Hooker (2)	Platte (4)
Dawes (7)	Jefferson (7)	Polk (1)
Dawson (3)	Johnson (2)	Red Willow (1)
Deuel (3)	Keith (2)	Richardson (3)
Dixon (1)	Kimball (1)	Sarpy (1)

Saunders (2)   Sheridan (4)   Thurston (1)

Scotts Bluff (4)   Sioux (8)   Webster (3)

Seward (1)   Thayer (3)

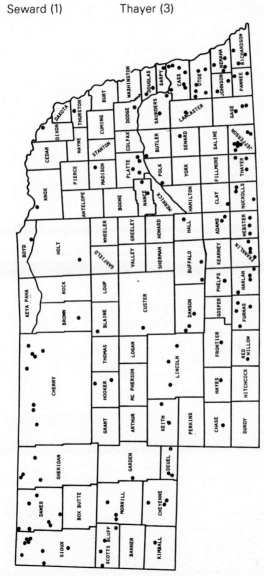

COLLECTING SITES IN NEBRASKA

(FOR SPECIFIC DETAILS ABOUT THESE
LOCATIONS, SEE THE FOLLOWING
COUNTY MAPS AND RELATED TEXT)

ADAMS COUNTY

1. Ayr area gravel pits.

     Agate. Jasper. Petrified wood. (10)

2. Ayr area: in the Little Blue River basin from the Crystal Lake Recreation Area downstream to the Clay County border.

     Agate. Jasper. Petrified wood. (1-3-7-18-20)

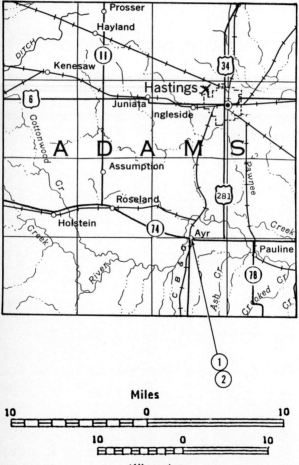

BLAINE COUNTY

1. Purdum area: collect along North Loup River upstream,
   along road running northwest from town through northeast corner of Thomas County and to area of Brownlee in
   Cherry County.

   Opal, wood. (1-7-12-20)

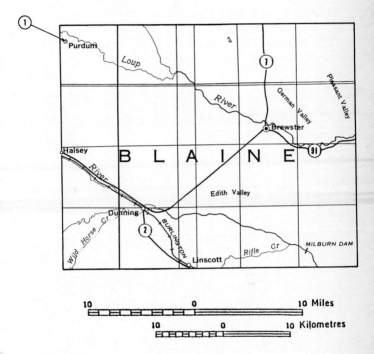

BOYD COUNTY

1. Spencer area: 4.5 miles south-southeast to Spencer Dam;
   collect upstream on both sides of the Niobrara River to
   SR-11.

   Fossils, fauna: ammonites; clams; fish; shark teeth.
   Fossils, flora: leaves of various trees. (4-5-Cretaceous
   limestone -7-14-20-22-29-42)

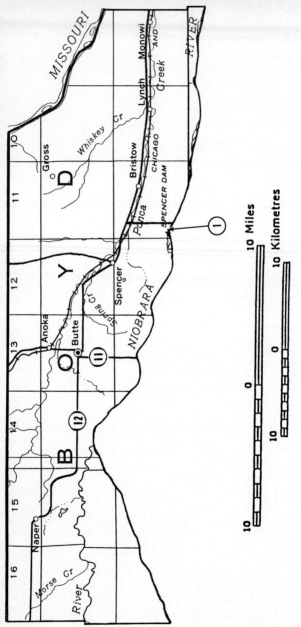

BROWN COUNTY

1. Ainsworth area: 20 miles south on US-7 to bridge over Calamus River; collect upstream on both sides for 5 miles, especially at the 5-mile point upstream.

Opal, wood. (1-7-13-20)

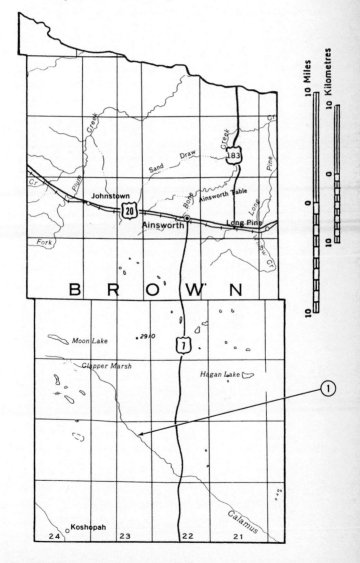

BUFFALO COUNTY

1. Elm Creek area: southwest to Platte River at Dawson County border; collect downstream on shorelines of both sides and in sand pits, all the way to the Hall County border in the southeast.

    Agate. Jasper. Petrified wood. (11-20)

2. Kearney area.

    Petrified wood. (10-11)

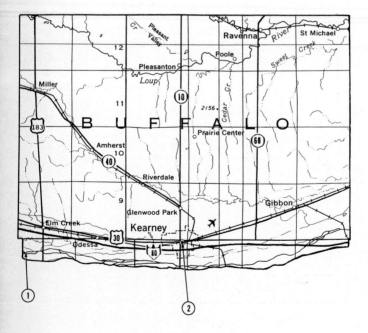

BUTLER COUNTY

1. Bellwood area: 6.9 miles west-northwest to the Platte River at the northwest corner of the county; collect downstream (east) along shorelines and in sand pits on both sides from the Polk County border in the northwest to the Saunders County border in the northeast.

   Agate. Jasper. Petrified wood. (11-20)

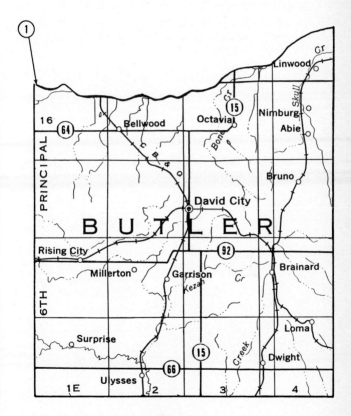

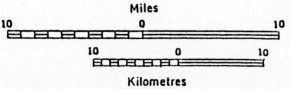

Miles

Kilometres

CASS COUNTY

1. Murray area: just off US-73/75; at the Queen Hill Quarry.

    Calcite: small crystals. Fossils, marine: chiefly inverte-
    brates; many silicified; good trilobites and excellent
    horn corals; occasional fish. Marcasite: crystals. Py-
    rite: brassy; bright crystals. Sphalerite. (12)

2. Nehawka area: northwest toward Weeping Water; at the
   Snyderville Quarry.

    Agate. Calcite: small crystals. Fossils: marine; chiefly
    invertebrates; especially brachiopods, crinoids, horn
    corals; many silicified. (12)

3. Plattsmouth area.

    Fossils, marine: chiefly horn corals; many silicified.
    (5-limestone -14)

4. Plattsmouth area: at the Onyx Ace Hill Quarry.

    Calcite (Onyx var.). (12)

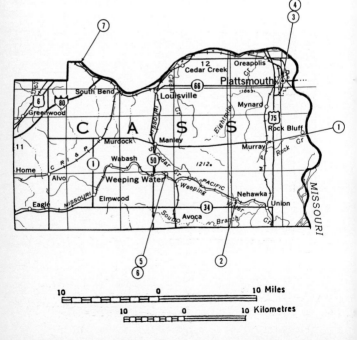

5. Weeping Water area around town and in quarries.

  Fossils: vertebrate. (1-12-13)

6. Weeping Water area: upstream and down, in and near Weeping Water Creek.

  Chert (Rice agate var.). (1-3)

7. South Bend area: 4.5 miles west-northwest to the Platte River at the Saunders County border; collect downstream (east) along shorelines and in sand pits on both sides, from the Saunders County border to the Platte River mouth at the Missouri River.

  Agate. Jasper. Petrified wood. (11-20)

CHASE COUNTY

1. Champion area: along both sides of Frenchman Creek; upstream to the Colorado border and downstream to the Enders Reservoir.

  Opal, wood. (1-4-7-13-20)

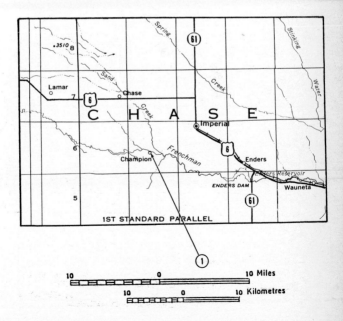

CHERRY COUNTY

1. Crookston area: upstream and down on both sides of Minnechaduza Creek and in the hills flanking it, all the way to the South Dakota border.

    Agate, fortification. Fossils: agatized and opalized; especially portions of mastodon tusks showing cross-section rings (which do not show in the petrified wood pieces). Petrified wood. (1-7-13-20)

2. Merriman area: upstream in Dry Creek and in hills flanking the creek, all the way west to the Sheridan County border.

    Agate, fortification. Fossils: agatized and opalized; occasional mastodon tusk sections. (1-7-13-20)

CHERRY COUNTY
NEBRASKA

DIVIDED INTO TWO
SEPARATE MAPS ON
THE FOLLOWING TWO
PAGES -- EACH MAP
IDENTIFIED BY ITS
NUMBER, AS SHOWN
IN DIAGRAM

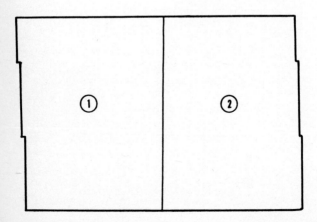

3. Merriman area: 0.8 mile east on US-20 to Dry Creek bridge; follow it 0.25 mile downstream to where it is joined by Bear Creek; then collect up Bear Creek and in the hills along both sides all the way north to the South Dakota border.

Agate, fortification. Fossils: agatized and opalized; some mastodon tusk cross sections showing rings (which are not present in the petrified wood). Jasper. Petrified wood. (1-7-13-20)

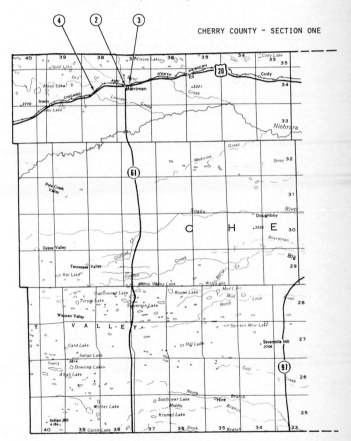

CHERRY COUNTY - SECTION ONE

4. Merriman area: 5.4 miles west on US-20 to Chicago & Northwestern RR bridge over Leander Creek; collect downstream in the creek and on the hills flanking it, all the way southeast to its mouth on the Niobrara River.

Agate, fortification. Fossils: agatized and opalized; some mastodon tusk cross sections showing rings. (1-7-13-20)

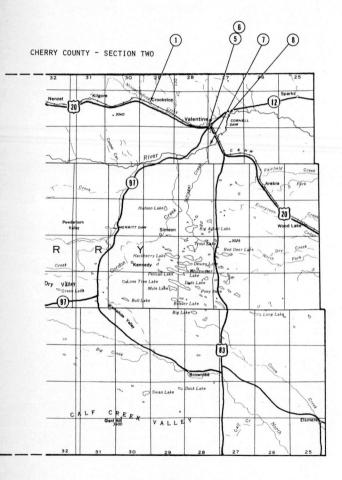

CHERRY COUNTY - SECTION TWO

5. Valentine area.

   Opal, wood. (1-4-13)

6. Valentine area: collect in Minnechaduza Creek and the hills flanking it, upstream to Dry Creek and downstream to SR-12.

   Agate: scarce. Jasper. Petrified wood: black. (1-7-13-20)

7. Valentine area: 3.1 miles south on US-20 to the Niobrara River; then upstream on both sides to the mouth of Schlagel Creek.

   Agate: some fortification. Fossils, flora and fauna: agatized and opalized; fine plant specimens; also cross sections of mastodon tusks showing rings (which do not show in the petrified wood). Petrified wood: black. (1-7-13-20)

8. Valentine area: 3.25 miles east on SR-12 to Cornell Dam; then downstream below dam in the hills flanking the Niobrara River.

   Petrified wood: black. (1-7-13-20)

CHEYENNE COUNTY

1. Gurley area: 1 mile south to Rush Creek bridge; collect upstream about 7.5 miles to the mouth of the second major tributary and downstream to the Garden County border.

   Agate. Chalcedony. Jasper. Petrified wood. (1-3-7-13-18-20)

2. Sidney area: along both sides of Lodgepole Creek, upstream (west-northwest) to the Kimball County border and downstream (east) to the Deuel County border.

   Agate. Jasper. Petrified wood. (1-7-11-13-18-20)

3. Sidney area: at the rest stop on the eastbound lane of I-80.

   Opal, common; some with color play; some dendritic; abundant. (4)

CLAY COUNTY

1. Deweese area: along both sides of the Little Blue River, upstream (northwest) to the Spring Ranch area and downstream (southeast) to the Nuckolls County border.

    Agate. Jasper. Petrified wood. (1-3-7-13-18-20)

CUSTER COUNTY

1. Arnold area: collect for 5 miles downstream along both sides of the South Loup River.

    Labradorite: some gem quality. (1-7-13-20)

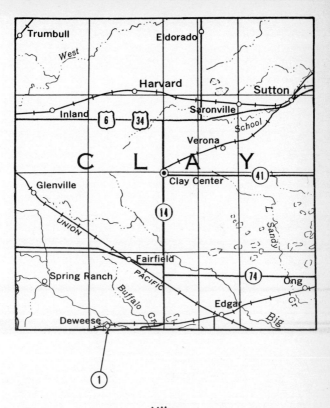

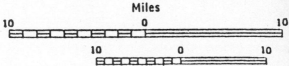

Miles

10   0   10

10   0   10

DAWES COUNTY

1. Chadron area: throughout the entire area north of US-20, from the Sheridan County border in the east to the Sioux County border in the west.

     Chalcedony: blue. (1-4-13)

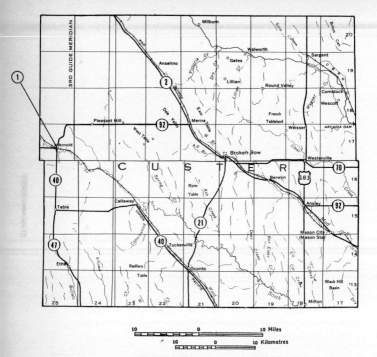

10 ——————— 0 ——————— 10 Miles

10 ——————— 0 ——————— 10 Kilometres

2. Chadron area: 2 miles west on US-20 to right turn on US-385; 4 miles north on US-385 to White River bridge; paying particular attention to concretions to 3″ diameter in which some of the material is found; collecting downstream (northeast) on both sides and on adjacent hills all the way northeast to the South Dakota border.

> Agate, fortification. Calcite: crystal clusters. Celestite: crystal clusters. Chalcedony: roses. Fossils, chiefly marine: silicified. Jasper. Petrified wood. (1-7-13-14-20-39-41)

3. Crawford area.

> Agate, fortification (Fairburn var.). (1-13-39)

4. Crawford area: just north; in breaks of the White River basin; in a series of extensive beds.

> Agate, blue. Agate, fortification (Fairburn var.). (1-13-39)

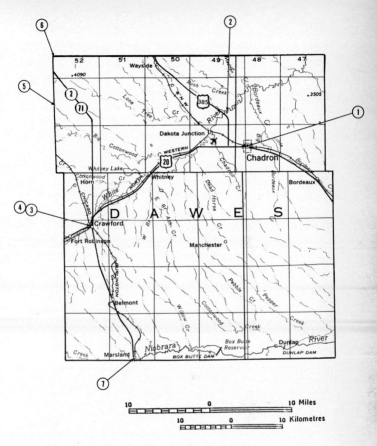

5. Crawford area: very extensive beds beginning 15 to 20 miles northwest and north-northwest; extending into Sioux County.

Agate, blue: geodes and nodules; some to 100 lb; excellent; some gem quality; best and largest found by digging; probing is a good method. Agate, fortification (Fairburn var.). (1-13-39)

6. Crawford area: 0.3 mile east to SR-2/71; 4.4 miles north to gravel road left; follow this 2 miles to where it turns north along east side of the Chicago, Burlington & Quincy RR; follow this north for another 11.7 miles to Orella Station in Sioux County; continue on same road another 0.6 mile to a gravel side road to the right; in the distance to the right is conspicuously isolated Sugarloaf Butte; drive 0.25 mile east to area of pale sand and clay, which marks the start of a roughly circular agate bed about 0.5 mile in diameter.

Agate, blue. Agate, red. (1-13-39)

7. Marsland area: south on SR-2/71 to the Niobrara River bridge; downstream on both sides to the Box Butte Reservoir.

Agate: fortification. Fossils, flora and fauna: many varieties of each; highly silicified; often entire skeletal fossils. (1-7-13-20-39)

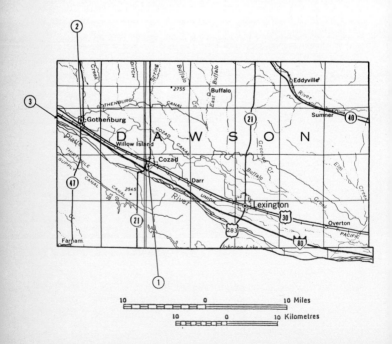

DAWSON COUNTY

1. Cozad area: in gravel pits and sand pits located east-south-east to the Overton area.

    Petrified wood. (10-11)

2. Gothenburg area sand pits.

    Petrified wood: palmwood. (11)

3. Gothenburg area: 3.1 miles west to Platte River at Lincoln County border; collect downstream along both shores and in sand pits all the way to the Buffalo County border in the southeast.

    Agate, fortification (Fairburn var.). Jasper. Petrified wood. (11-18-20)

DEUEL COUNTY

1. Big Springs area: upstream (southwest) along both sides of the South Platte River and its immediate basin to the Colorado border.

    Agate. Jasper: red, yellow. Petrified wood. (1-7-11-13-18-20)

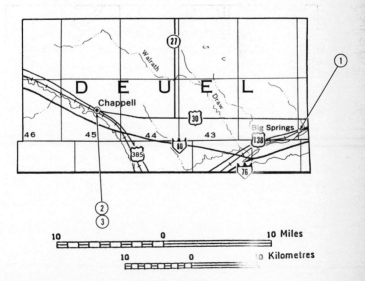

2. Chappell area: collect on hills, especially during spring in newly plowed wheat fields in a wide radius around town.

   Agate. Chalcedony. Jasper. Petrified wood: cottonwood; cycads, palm. Quartz, cat's-eye. Quartz, rock. (1-4-13)

3. Chappell area: along Lodgepole Creek on both sides and in flanking hills upstream (northwest) to the Cheyenne County border and downstream (southeast) to the Colorado border.

   Agate. Chalcedony. Jasper: red, yellow. Petrified wood: some palmwood. (1-3-7-13-20)

DIXON COUNTY

1. Ponca area: 5.2 miles north; collect along the Missouri River; in Ponca State Park.

   Fossils, fauna: ammonites; clams; fish; shark teeth. Fossils, flora: leaves of various trees. (4-5-Cretaceous limestone -7-14-20-22-29-42)

DODGE COUNTY

1. Fremont area.

   Agate, fortification. (1-13)

DOUGLAS COUNTY

1. Valley area.

   Agate, common. Agate, moss. Chert. Opal, moss. Opal, wood. (3)

2. Waterloo area: southwest; in the Lyman-Ritchie Gravel Pit.

   Agate. (10)

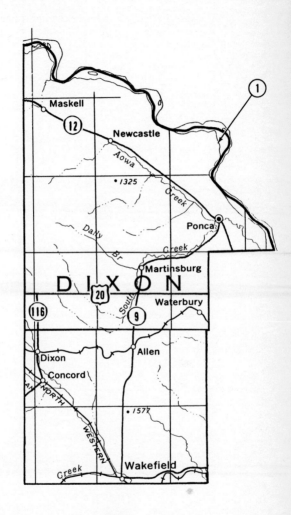

Maskell

Newcastle

Ponca

Martinsburg

Waterbury

Allen

Dixon

Concord

Wakefield

D I X O N

• 1325

• 1577

Aowa Creek

Daly Br

Creek

South

NORTH WESTERN

Creek

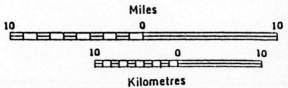

Miles

Kilometres

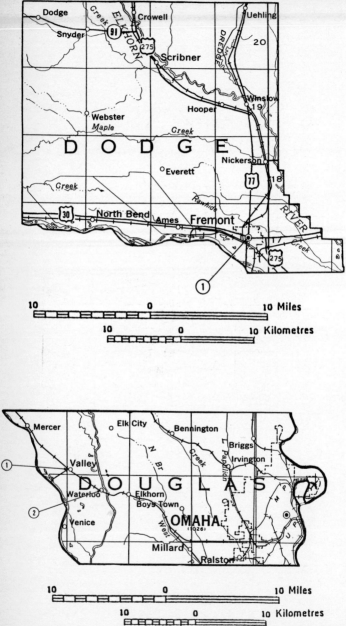

10 Miles

10 Kilometres

10 Miles

10 Kilometres

FRANKLIN COUNTY

1. Bloomington area: 1.5 miles south; in exposures along the Republican River, especially the south side.

    Fossils, marine: ammonites; clams; fish; oysters; shark teeth. (4-5-Cretaceous limestone -7-20)

2. Franklin area: in exposures south of town, especially along the south side of the Republican River.

    Fossils, marine: ammonites; clams; fish; oysters; shark teeth. (4-5-Cretaceous limestone -7-20)

3. Franklin area: in the Republican River basin upstream to the Naponee area.

    Jasper: pastel colors. (1-3-18-20)

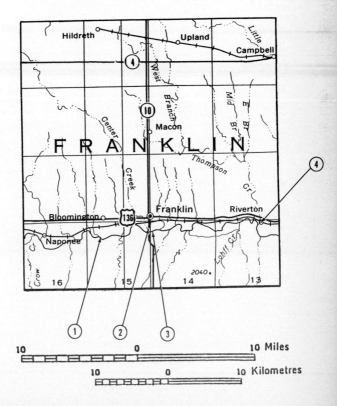

4. Riverton area: in exposures along the Republican River, downstream to the Webster County border and upstream for 7 miles.

> Fossils, flora: variety of tree leaves. Fossils, marine: ammonites; clams; fish; shark teeth. (4-5-Cretaceous limestone -7-14-20-22-29-42)

FRONTIER COUNTY

1. Stockville area: along both sides of Medicine Creek, downstream (southeast) to Harry Strunk Lake and upstream (northwest) to the Maywood area.

> Opal, wood. (1-4-7-13-20)

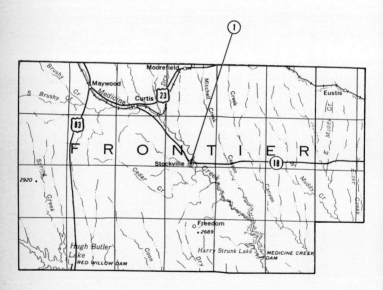

FURNAS COUNTY

1. Arapahoe area.

   Quartz, smoky: some gem quality. (1-4-13)

2. Beaver City area: in Niobrara chalk beds.

   Chert: green, red. (5-chalk -14-39)

3. Cambridge area: in the Republican River basin downstream to the Oxford area.

   Jasper: pastel colors. (1-3-18-20)

4. Edison area: 0.6 mile south to bridge over Republican River; collect along south side of river downstream to the Harlan County border.

   Fossils, flora: various leaves. Fossils, marine: ammonites; clams; fish; oysters; shark teeth. (4-5-Cretaceous limestone -7-20)

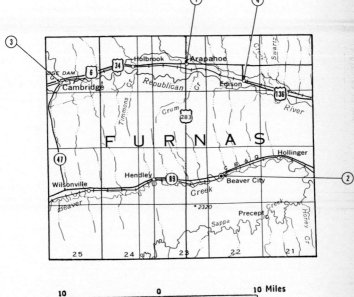

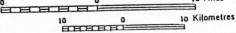

GAGE COUNTY

1. Barneston area: 3 miles north-northwest; on east bank of the Big Blue River.

   Celestite: blue. Quartz: geodes to 5″ diameter; some with interior celestite crystal linings. (12)

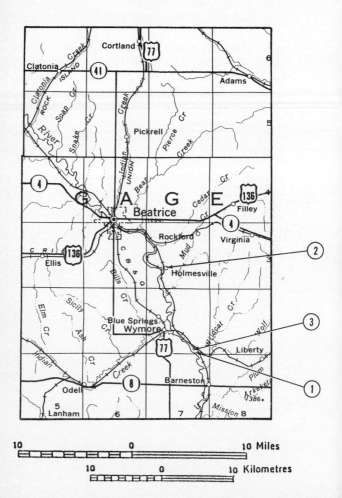

2. Holmesville area: just southwest; at quarry adjacent to west side of the Big Blue River; in a stratum abundant with geodes.

> Celestite: blue. Fossils, marine: invertebrates; silicified. Quartz: geodes to 5″ diameter; often with celestite crystal linings. (12)

3. Wymore area: 3 miles southeast; in the Gage County Road Department quarry near the Big Blue River.

> Celestite: blue. Quartz: geodes to 5″ diameter; some including fine blue celestite crystals. (12)

GARDEN COUNTY

1. Lisco area: along both banks of the Platte River, from the water's edge to several hundred yards outward and also in sand pits, upstream to the Morrill County border and downstream to the Keith County border.

> Agate. Jasper. Opal, wood: gem quality. Petrified wood. (4-13-11-20)

HALL COUNTY

1. Grand Island area: collect upstream along both sides of the Platte River and in adjacent sand pits from the Hamilton County border to the Buffalo County border.

> Agate. Jasper. Petrified wood. (4-11-20)

HARLAN COUNTY

1. Alma area.

> Opal, wood. (1-4-13)

2. Alma area: in exposure south of town; especially along south bank of Harlan County Lake.

> Fossils, marine: ammonites; clams; fish; oysters; shark teeth. (4-5-Cretaceous limestone -7-20)

10   0   10 Miles

10   0   10 Kilometres

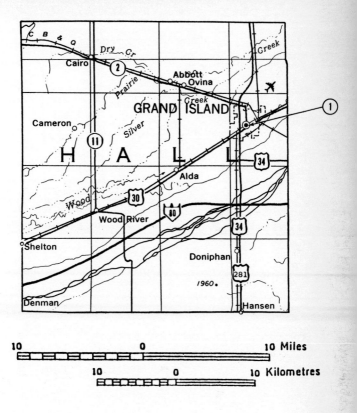

3. Oxford area: downstream (southeast) along both sides of the Republican River to the Orleans area.

   Jasper: pastel colors. (1-3-7-18-20)

4. Republican City area: from the Harlan County Lake Dam downstream along both sides of the Republican River to the Franklin County border; also some reported from fill material below Harlan County Lake Dam.

   Fossils, marine: ammonites; clams; fish; oysters; shark teeth. Jasper: pastel colors. (1-3-4-7-18-20-37)

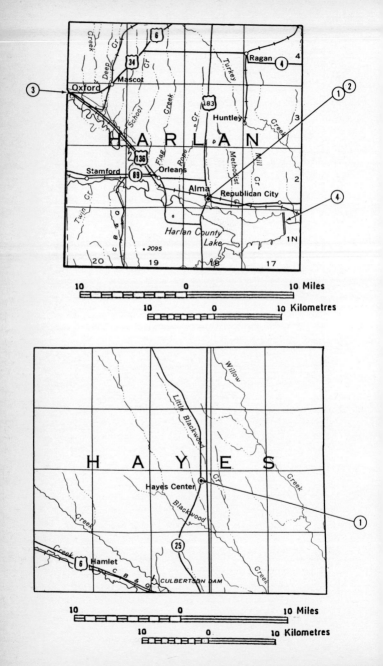

HAYES COUNTY

1. Hayes Center area: east and northeast on the Maywood Road to Red Willow Creek; collect upstream and down along both sides for several miles.

    Opal, wood. (1-4-7-13-20)

HOLT COUNTY

1. Swan Lake area: 0.8 mile north to South Fork of the Elkhorn River; collect upstream (west) along both sides to headwaters and downstream (east-southeast) for 5 miles.

    Opal, wood. (1-3-4-7-13-20)

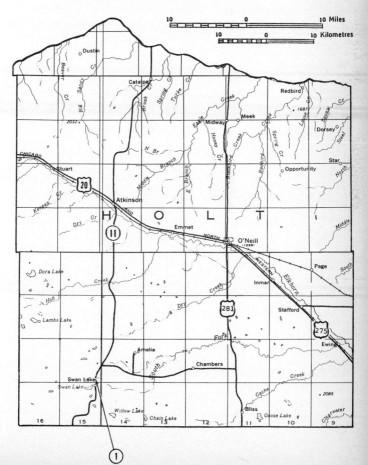

HOOKER COUNTY

1. Mullen area: 1.6 miles north on SR-97 to bridge over Middle Loup River; collect along both sides upstream (northwest) to the Cherry County border and downstream (east) to the Thomas County border.

    Labradorite: some gem quality. (1-7-13-20)

2. Mullen area: 12.2 miles south on SR-97 to bridge over Dismal Creek; go upstream 1 mile on south bank to South Fork entering from southwest; collect upstream on South Fork along both shores to Jefford Lake.

    Agate: sparse. Jasper, yellow. Opal, wood: some fine gem quality. (1-4-7-13-20)

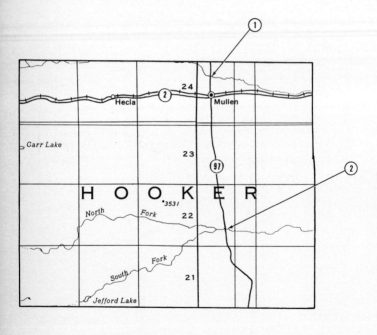

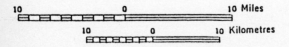

JEFFERSON COUNTY

1. Fairbury area.
    Agate. Jasper. Petrified wood. (1-3-13)

2. Fairbury area: in road cuts and along the Little Blue River.
    Fossils, flora: fine leaf fossils of various trees.
    (1-4-7-13-14-20-22-28-29-42)

3. Fairbury area gravel pits.

> Agate. Jasper. Petrified wood. (10)

4. Fairbury area: south on SR-15 to near Kansas border; especially in road cuts, stream cuts and outcroppings.

> Fossils, marine: ammonites; clams; fish; oysters; shark teeth. (1-4-5-Cretaceous limestone -7-13-14-20-22-28-29-42)

5. Powell area: along both sides of the Little Blue River upstream (northwest) to the Thayer County border and downstream (southeast) to the Kansas border.

> Agate. Jasper. Petrified wood. (1-3-7-13-20)

6. Steele City area.

> Agate. Jasper. Petrified wood. (1-13)

7. Steele City area gravel pits.

> Agate. Jasper. Petrified wood. (10)

JOHNSON COUNTY

1. Elk Creek area: in gravel exposures along the North Fork of the Big Nemaha River.

> Agate, fortification (Lake Superior var.). (1-3-7-10-13-18)

2. Sterling area.

> Agate, fortification (Lake Superior var.). (1-3-7-10-13-18)

KEITH COUNTY

1. Ogallala area: upstream (west) along the north bank of the South Platte River to the Deuel County border.

> Agate. Jasper: red, yellow. Petrified wood. (1-11-13-20)

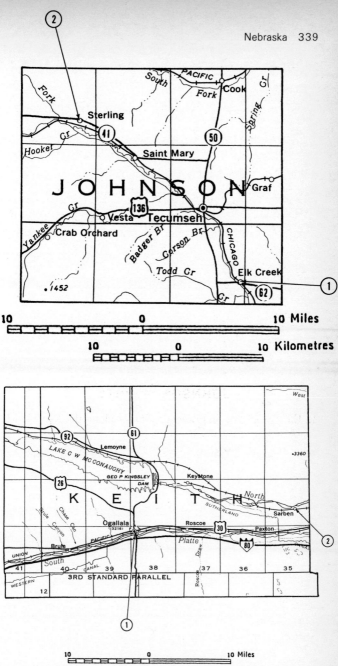

10   0   10 Miles

10   0   10 Kilometres

10   0   10 Miles

10   0   10 Kilometres

2. Sarben area: collect upstream (west) along the south bank of the North Platte River to the Kingsley Dam; also in adjacent sand pits.

> Agate. Jasper. Petrified wood. (1-4-11-20)

KIMBALL COUNTY

1. Kimball area: collect in road cuts and exposures south of town all the way to the Colorado border.

> Opal, common: some with color play; some dendritic.
> Opal, wood: some gem quality. (1-4-13)

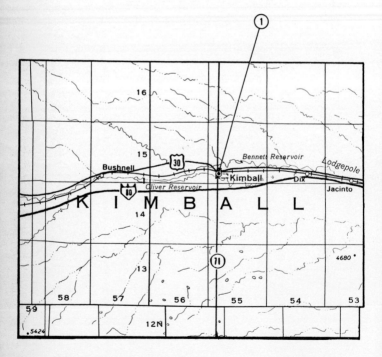

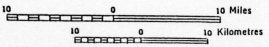

KNOX COUNTY

1. Niobrara area: in exposures surrounding town, especially in road cuts and along the Niobrara River.

   Fossils, flora: leaves of various trees, including hardwoods. Fossils, marine: ammonites; clams; fish; oysters; shark teeth. (4-5-Cretaceous limestone -7-14-20-22-29-42)

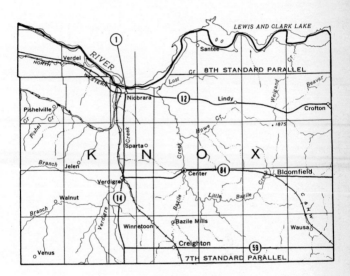

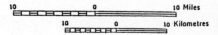

LANCASTER COUNTY

1. Lincoln area.

   Agate, fortification. Chert (Rice agate var.). (1-3-10-12)

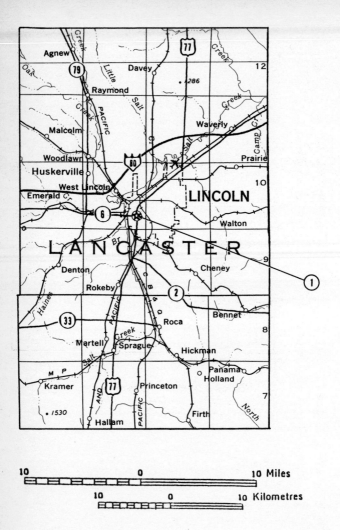

LINCOLN COUNTY

1. Hershey area: upstream in the Platte River basin (aka the Land Between the Rivers) and in area sand pits to the Keith County border.

   Agate. Jasper: red, yellow. Petrified wood. (1-11-13-20)

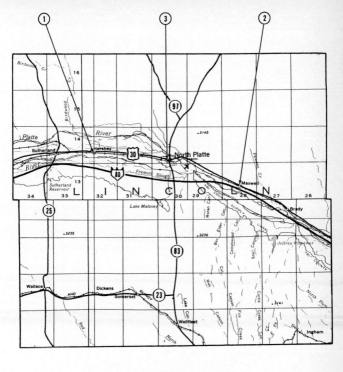

2. Maxwell area: downstream in the Platte River basin to the
   Dawson County border; also in adjacent sand pits.
     Agate. Jasper: red, yellow. Petrified wood. (1-11-13-20)

3. North Platte area gravel pits.
     Agate. Jasper: red, yellow. Petrified wood. (10)

MADISON COUNTY

1. Norfolk area gravel pits.
     Agate, fortification. (10)

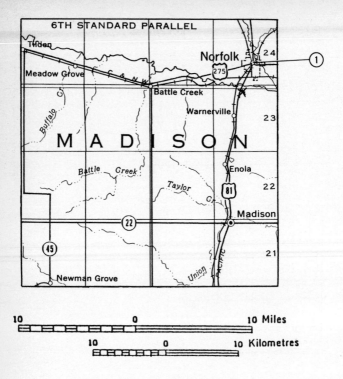

MORRILL COUNTY

1. Angora area (aka Angora Hill); 2 miles east on faint road; then left on another faint road north up hill to parking area: continue on foot to crest of hill; in and around abandoned mine.

    Opal, moss: white opaque to gray translucent; also a green that fluoresces; in chalky-coated nodules. (1-2-5- limestone -13)

2. Bayard area.

    Chert, banded: black or gray exterior; interior concentrically banded with yellow, brown, red, orange, and mixed colors. (3-26)

3. Bayard area: south on US-26 to the North Platte River;
   then along both shorelines and in adjacent sand pits downstream to the Garden County border.

   Agate. Jasper. Petrified wood. (1-4-11-13-20)

NANCE COUNTY

1. Fullerton area: widespread in a radius of about 10 miles.
   Agate. Petrified wood. (10-39)

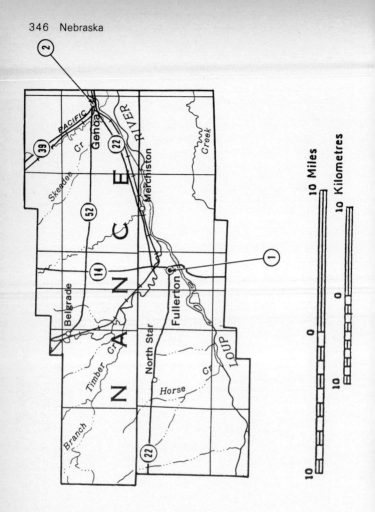

2. Genoa area: upstream on both sides of the Loup River from the Platte County border to the Fullerton area.

   Agate. Petrified wood. (1-3-13-18-20)

NEMAHA COUNTY

1. Auburn area.

   Agate, fortification (Lake Superior var.). (1-3-7-10-13-18)

2. Brock area.
   Agate, fortification (Lake Superior var.). (1-3-7-10-13-18)

3. Johnson area.
   Agate, fortification (Lake Superior var.). (1-3-7-10-13-18)

4. Johnson area: downstream in the small creek running north to the Little Nemaha River near Brock.
   Petrified wood. (3-7-13-20)

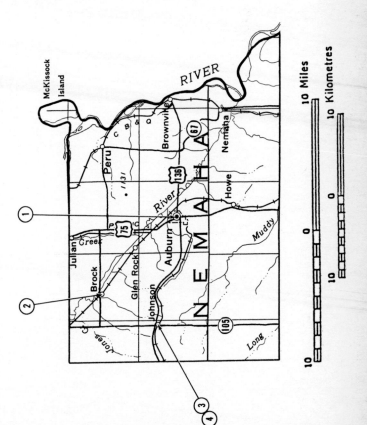

NUCKOLLS COUNTY

1. Oak area gravel pits.

    Agate, fortification (Lake Superior var.). (10)

2. Oak area: along both sides of the Little Blue River upstream to the Clay County border and downstream to the
   Thayer County border.

    Agate. Jasper. Petrified wood. (1-3-18-20)

3. Superior area: along the south side of the Republican
   River, upstream to the Webster County border and downstream to the Kansas border.

    Fossils, marine: ammonites; clams; fish; oysters;
    shark teeth. (4-5-Cretaceous limestone -7-20)

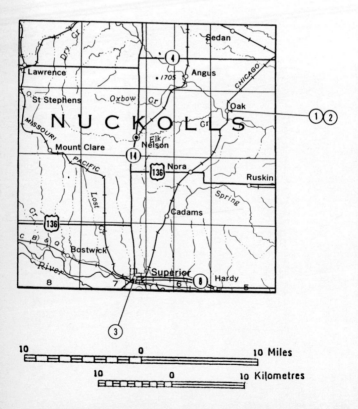

OTOE COUNTY

1. Douglas area gravel deposits.
   Agate, fortification (Lake Superior var.). (1-3-7-10-13-18)

2. Dunbar area gravel deposits.
   Agate, fortification (Lake Superior var.). Agate, moss. (1-3-7-10-13-18)

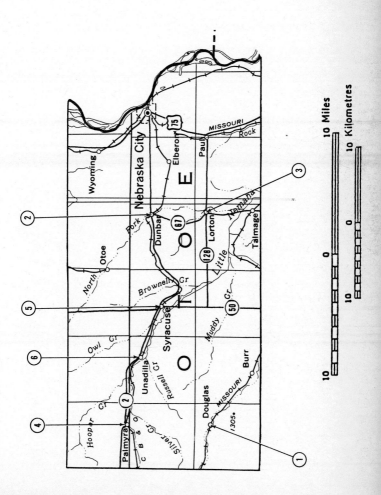

3. Lorton area gravel deposits.

   Agate, fortification (Lake Superior var.). (1-3-7-10-13-18)

4. Palmyra area gravel deposits.

   Agate, fortification (Lake Superior var.). Agate, moss. (1-3-7-10-13-18)

5. Syracuse area gravel deposits.

   Agate, fortification (Lake Superior var.). (1-3-7-10-13-18)

6. Unadilla area gravel deposits.

   Agate, fortification (Lake Superior var.). (1-3-7-10-13-18)

PAWNEE COUNTY

1. Du Bois area gravel deposits.

   Agate, fortification (Lake Superior var.). (1-3-7-10-13-18)

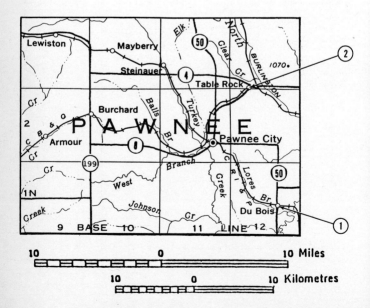

2. Table Rock area gravel deposits, especially in the vicinity
   of abandoned commercial gravel works.

   > Agate, fortification (Lake Superior var.). Jade.
   > Quartz, cat's-eye. Quartz, rock: some rutilated; some
   > tourmalinated. Quartz, smoky: asteriated. Unakite.
   > (1-3-7-10-13-18)

PHELPS COUNTY

1. Holdrege area and northward throughout the northern
   half of the county.

   > Petrified wood. (10-11)

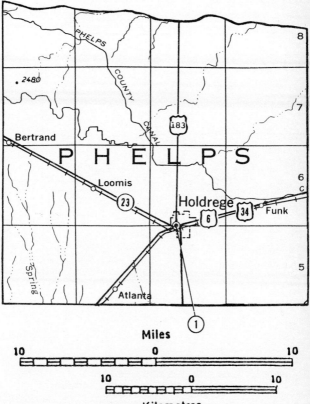

Miles

Kilometres

PLATTE COUNTY

1. Columbus area gravel pits.

    Jasper. Petrified wood. (10)

2. Columbus area: where the Loup River meets the Platte River.

    Jasper. Petrified wood. (1-3-18-20)

3. Oconee area: along both sides of the Loup River, upstream to the Nance County border and downstream to the Columbus area.

    Jasper. Petrified wood. (1-3-18-20)

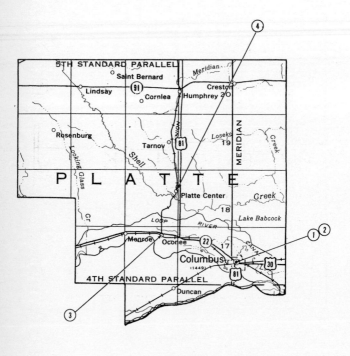

4. Platte Center area.

      Jasper. Petrified wood. (1-3-18-20)

POLK COUNTY

1. Osceola area: 13.3 miles west on SR-92 to the Platte River
   at the Merrick County border; collect downstream along
   both shorelines and in adjacent sand pits all the way to the
   Butler County border in the northeast.

         Agate, fortification (Lake Superior var.). Jasper. Pet-
   rified wood. (11-20)

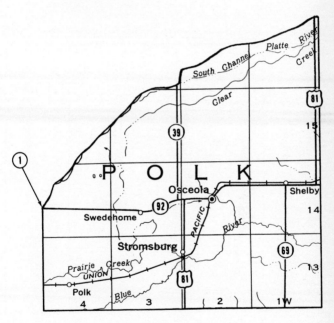

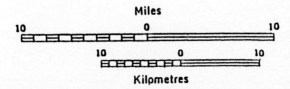

RED WILLOW COUNTY

1. **McCook area:** 1 mile south on US-83 to the bridge over the Republican River; collect downstream along the south shore of the Republican River to the Furnas County border and downstream along the north shore to the mouth of Red Willow Creek.

   Agate: sparse. Fossils, marine: ammonites; clams; fish; oysters; shark teeth. Jasper: pastel colors. (1-3-4-5-Cretaceous limestone -7-18-20)

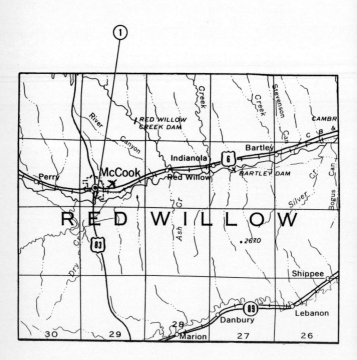

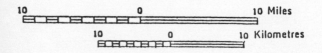

RICHARDSON COUNTY

1. Dawson area.

    Agate, fortification (Lake Superior var.). (1-3-7-10-13-18)

2. Humboldt area: along the Big Nemaha River, upstream to the Pawnee County border and downstream to its mouth at the Missouri River.

    Agate, fortification (Lake Superior var.). Quartz, smoky; gem quality (one fine piece found close to Humboldt was cut to a beautiful 1.55-ct. gem). (1-3-7-10-13-18-20)

3. Stella area gravel deposits.

    Agate, fortification (Lake Superior var.). (1-3-7-10-13-18)

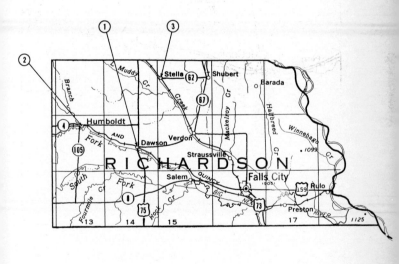

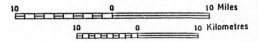

SARPY COUNTY

1. La Platte area: in sand pits along the Platte River, upstream to the Saunders County border.

   Agate. Petrified wood. (11)

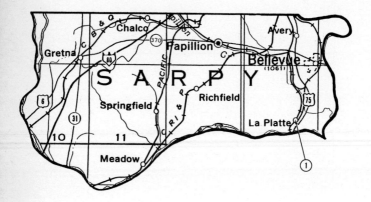

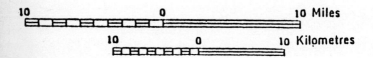

SAUNDERS COUNTY

1. Ashland area: 2.6 miles northeast on US-6 to the Platte
   River; collect upstream along both shorelines and in adjacent sand pits all the way to the Butler County border in
   the northwest.

   Agate, fortification (Lake Superior var.). (11-20)

2. Wahoo area gravel pits.

   Agate, fortification (Lake Superior var.). (10)

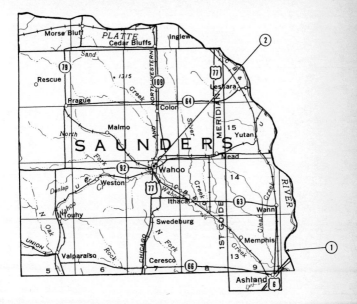

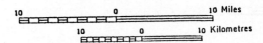

SCOTTS BLUFF COUNTY

1. Lyman area: north on SR-L79C to the Platte River; collect along both sides of the river and in adjacent sand pits.

    Amethyst, fortification (Lake Superior var.). Jasper. Petrified wood. (11-20)

2. McGrew area: upstream along the North Platte River to Terrytown.

    Agate. (3-18-20)

3. Mitchell area: upstream along the North Platte River to the South Morrill area.

Agate. Petrified wood: some dendritic. (3-18-20)

4. Scotts Bluff area road cuts; especially northward on SR-71.
Opal, wood: pure white. (4)

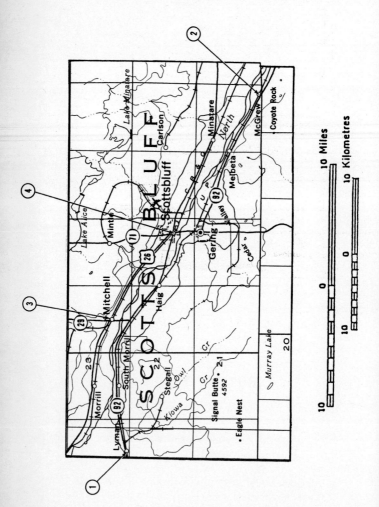

SEWARD COUNTY

1. Garland area gravel pit.
   Agate, fortification. (10)

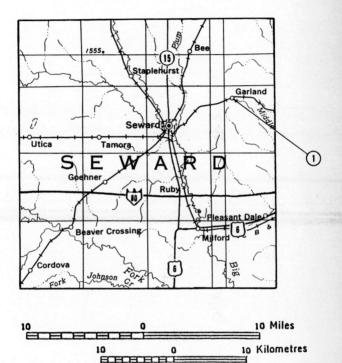

SHERIDAN COUNTY

1. Gordon area sand pits.
   Agate, fortification (Fairburn var.). (11)

2. Hay Springs area.
   Agate, fortification (Fairburn var.). Chalcedony. Jasper. Petrified wood. (3-4)

3. Whiteclay area: in broad region of beds to the Wyoming
   border.

   Agate, fortification (Fairburn var.). Chalcedony. Fossils, marine. Jasper. Opal, wood. Petrified wood.
   (1-3-4-13-14-39)

4. Whiteclay area: in gravel bars of White Clay Creek.

 Agate, fortification (Fairburn var.). (18)

SIOUX COUNTY

1. Agate area: collect in area adjacent to Agate Fossil Beds National Monument, especially upstream on the Niobara River.

 Agate, fortification (Fairburn var.). Calcite: crystals. Celestite: clusters of blue crystals. Fossils, flora and fauna: some of the world's finest, including whole preserved skeletons of many animal species, as well as fine plant specimens. (1-5-sandstone 7-13-14-20-39)

2. Harrison area: north on SR-29 (aka the Ardmore Road) to the Oglala National Grassland, 4.5 miles south of the South Dakota border; collect westward for several miles; fair material on surface; best material is subsurface, so probing is an effective method of locating specimens.

 Agate, blue. Agate, fortification (Fairburn var.). (1-13-39)

3. Harrison area: collect northward along both sides of SR-29 (aka the Ardmore Road) all the way to the South Dakota border, paying particular attention to road cuts, washes, draws, and streambeds; best material found within concretions to 10″ diameter.

 Agate, fortification. Calcite: crystals. Celestite: blue; crystal clusters. Chalcedony: roses. Fossils, marine (chiefly); many silicified. (1-13)

4. Harrison area: 8 miles west to the last side trail south before reaching the Wyoming border; go south on this trail to the Niobrara River; collect downstream from there all the way to the west boundary of Agate Fossil Beds National Monument.

 Agate, blue. Agate, fortification (Fairburn var.). Fossils: especially marine fauna. (1-3-4-7-13-18-20-29)

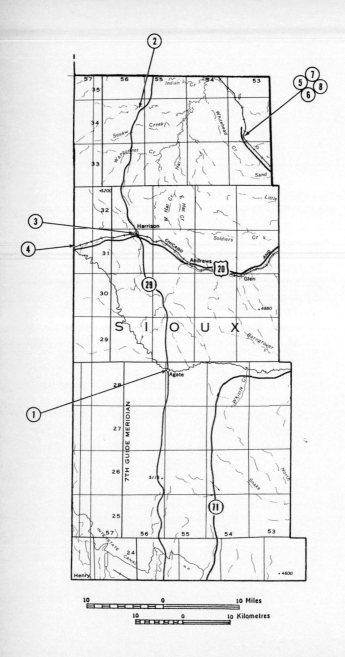

5. Orella Station area (a former watering stop of the Chicago, Burlington & Quincy RR located 20.5 miles northwest of Crawford in adjoining Dawes County and 5 miles south of Ardmore in Fall River County, South Dakota).

   Agate (Fairburn var.). (13-39)

6. Orella Station area (a former watering stop of the Chicago, Burlington & Quincy RR located 20.5 miles northwest of Crawford in adjoining Dawes County and 5 miles south of Ardmore in Fall River County, South Dakota): 0.5 mile south; then west across tracks and continue 2 miles; collect in Badlands.

   Agate (Fairburn var.). Jasper. Petrified wood. (1-13-39)

7. Orella Station area (a former watering stop of the Chicago, Burlington & Quincy RR located 20.5 miles northwest of Crawford in adjoining Dawes County and 5 miles south of Ardmore in Fall River County, South Dakota): 0.5 mile south on SR-29 to right turn on faint dirt road; cross tracks and go 3 miles west through Badlands area (negotiable for autos but four-wheel drive is recommended) to a region of canyons; collect in canyon walls and below them.

   Agate, fortification (Fairburn var.). Fossils, marine: some gastropods and corals; some silicification. Fossils, vertebrate: dinosaur and mammalian bone; often well silicified; mammalian bones and frequent teeth. Jasper: red, yellow. Petrified wood. Opal, wood: some of excellent quality. (1-5-sandstone -8-13-14-39)

8. Orella Station area (a former watering stop of the Chicago, Burlington & Quincy RR located 20.5 miles north of Crawford in adjoining Dawes County and 5 miles south of Ardmore in Fall River County, South Dakota): 1 mile north to ranch road right at gate; follow ranch road 2 miles northwest to Bald Butte; in a wide area surrounding this solitary butte.

   Agate, fortification. Carnelian. Chalcedony: geodes to 8″ diameter with interiors of fine agate and bright carnelian. Fossils, flora: well silicified ferns (including *Tempskya superba* as well as palmwood, teredo wood, pteriosperm, and cycads). Jasper: nodules to 7″ diameter; black exterior, cherry red interior; some with pockets lined with blue-gray opal. Selenite: enclosed in chalcedony. (1-13-39-40-41)

THAYER COUNTY

1. Alexandria area: in road cuts and other exposures, especially south of town in road cuts of SR-53.

   Fossils, flora: leaves of various trees, including hardwoods. Fossils, marine: ammonites; clams; fish; shark teeth. (4-5-Cretaceous limestone -7-14-20-22-29-42)

2. Gilead area: in road cuts and other exposures.

   Fossils, flora: leaves of various trees, including hardwoods. Fossils, marine: ammonites; clams; fish; shark teeth. (4-5-Cretaceous limestone -7-14-20-22-29-42)

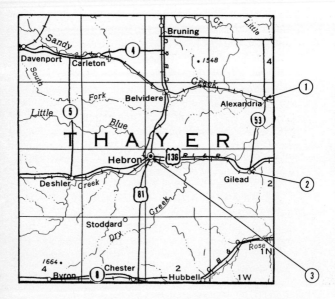

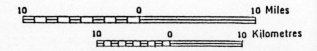

3. Hebron area: along both sides of the Little Blue River upstream to the Nuckolls County border and especially downstream to the Jefferson County border.

> Agate. Fossils, flora: plants and leaves, including hardwoods. Fossils, marine: ammonites; belemites; clams; fish; oysters; shark teeth. Jasper. Petrified wood. (1-3-18-19)

THURSTON COUNTY

1. Winnebago area gravel pit.

> Agate, fortification. (10)

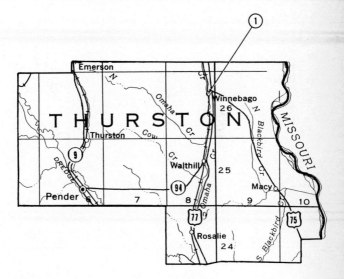

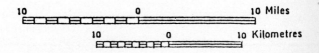

WEBSTER COUNTY

1. Guide Rock area: in exposures to the south, especially along the south side of the Republican River.

    Fossils, marine: ammonites; clams; fish; oysters; shark teeth. (4-5-Cretaceous limestone -7-18-20)

2. Inavale area: in exposures along the Republican River upstream to the Franklin County border and downstream to the Superior-Courtland Dam.

    Fossils, flora: plants; leaves of various trees, including conifers and hardwoods. Fossils, marine: ammonites; clams; fish; oysters; shark teeth. (4-5-Cretaceous limestone -7-18-20)

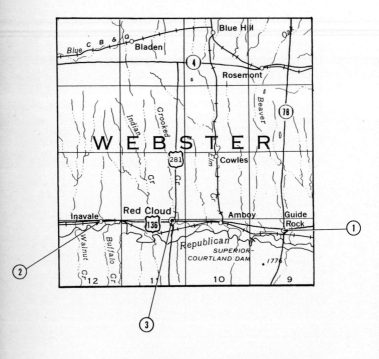

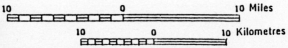

3. Red Cloud area: exposures and cuts south of town.

Fossils, marine: ammonites; clams; fish; oysters; shark teeth. (4-5-Cretaceous limestone -7-18-20)

NORTH DAKOTA

Statewide total of 52 locations in the following counties:

Adams (1)	Kidder (2)	Ramsey (5)
Billings (5)	McHenry (1)	Ransom (2)
Burleigh (2)	McKenzie (4)	Rolette (1)
Dunn (1)	Mercer (1)	Slope (2)
Golden Valley (2)	Morton (1)	Stark (5)
Grant (2)	Mountrail (2)	Ward (1)
Hettinger (7)	Pembina (3)	Williams (2)

COLLECTING SITES IN NORTH DAKOTA

(FOR SPECIFIC DETAILS ON THESE
LOCATIONS, SEE THE FOLLOWING
COUNTY MAPS AND RELATED TEXT)

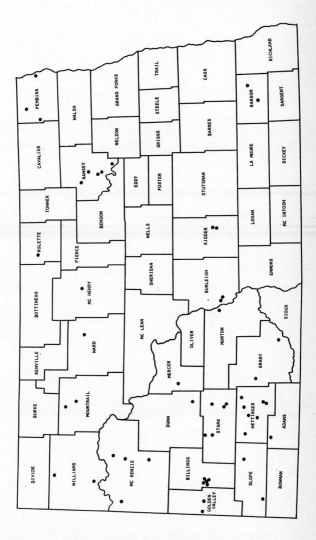

ADAMS COUNTY

1. Reeder area: 3 miles east on US-12 to SR-22; 9.9 miles north on SR-22 to the Cedar River; collect downstream southeast on both sides and in adjacent hills all the way to the Grant County border.

    Petrified wood. (1-7-13-20)

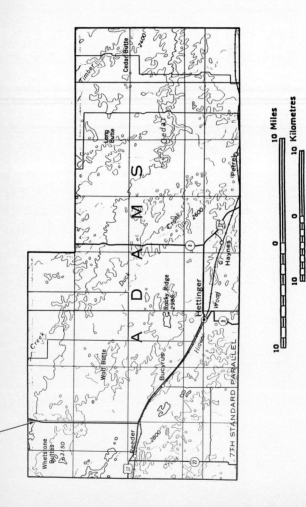

BILLINGS COUNTY

1. Medora area: in surrounding Badlands.
   Petrified wood. (1)

2. Medora area: in area of eroded buttes, cliffs, and pinnacles in Badlands.
   Aragonite. Barite: occasional. Calcite: yellow crystals. Siderite. (1-13-14-41-all)

3. Medora area: in canyons to the south.
   Petrified wood. (1-13)

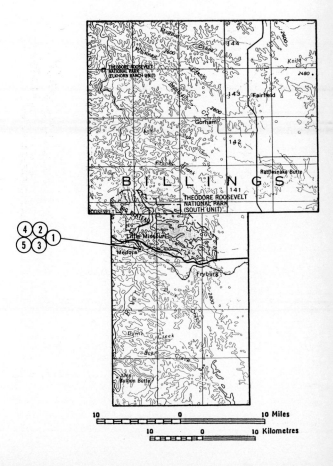

4. Medora area: northward along the east bank of the Little Missouri River.

   Petrified wood. (1-3-13-18-20)

5. Medora area: southward along both sides of the Little Missouri River.

   Calcite: good crystals. Fossils, fauna and flora: many centered in concretions to 16″ diameter. (1-13-14-29-41)

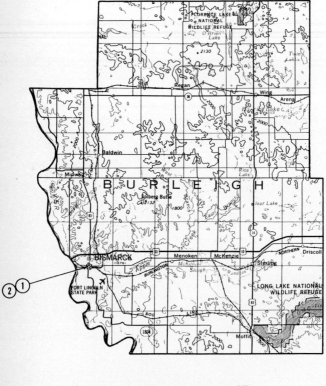

BURLEIGH COUNTY

1. Bismarck area gravel pits along the Missouri River.
   Agate, fortification. (10)

2. Bismarck area: in the Cannonball Formation.
   Petrified wood. (1-13)

DUNN COUNTY

1. Marshall area: 13 miles southwest; 10 miles north of Taylor (Stark County); 3.4 miles north of the Stark County border.
   Petrified wood. (1-13-39)

GOLDEN VALLEY COUNTY

1. Beach area: to the Lutheran Church area mines and prospects.
   Uranium: with associated minerals. (2-21)

2. Sentinel Butte area mines and prospects.
   Uranium: with associated minerals. (2-21)

GRANT COUNTY

1. New Leipzig area: southwest to the Cannonball River at the Hettinger County border; then downstream southeast on both sides to the mouth of the Cedar River, 1.7 miles east of SR-31.
   Fossils, flora: *Katsura* tree pods; osmundite plant matter; sequoia cones to walnut size; *Tempskya* ferns; all well silicified. Petrified wood. (1-3-7-13-20)

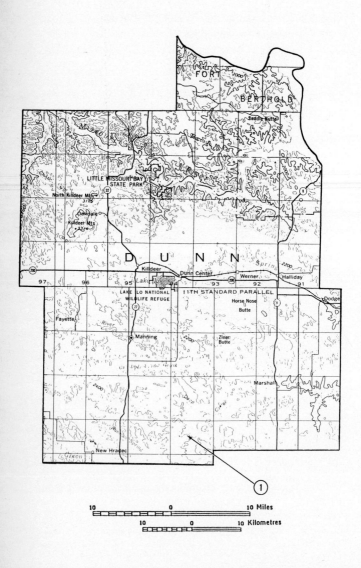

10   0   10 Miles

10   0   10 Kilometres

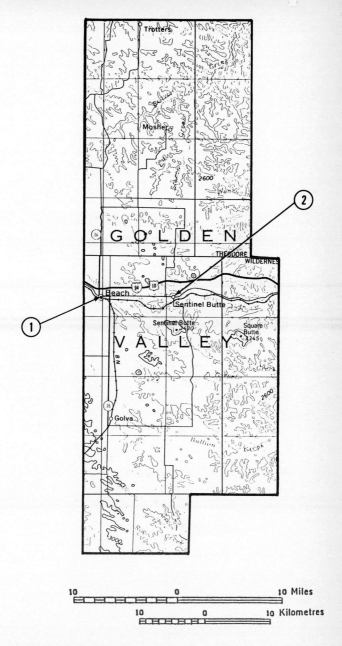

2. Raleigh area: 13.3 miles south on SR-31 to bridge over
   Cannonball River; downstream to mouth of Cedar River;
   collect in triangular area between the two rivers just
   above where they meet.

   Fossils, flora: sequoia cones to walnut size; well silicified. Petrified wood. (1-13-18-20)

HETTINGER COUNTY

1. Mott area: especially in Cannonball River valley.
   Selenite: fine crystals. (1-7-13-20)

2. Mott area: north along both sides of SR-8 to the Stark
   County border.
   Agate, fortification. Jasper. Petrified wood. (1-13)

3. Mott area: 11 miles north; on hills just southwest of Camel Buttes and also on the Camel Buttes.

   Agate, fortification. Chalcedony. Jasper. (1-3-13-26)

4. New England area: upstream on both sides of the Cannonball River to the Slope County border.

   Agate, fortification. Petrified wood. (1-3-7-13-20)

5. Regent area: just north; in the hills flanking the Cannonball River.

   Petrified wood. (1-13-20)

6. Regent area: 2 miles northeast; then downstream along both sides of the tributary of the Cannonball River and in adjacent hills to the mouth of this stream at the Cannonball River.

   Agate, fortification. Petrified wood. (1-7-13-20)

7. Regent area: 6 miles northeast.

   Petrified wood. (1-13-39)

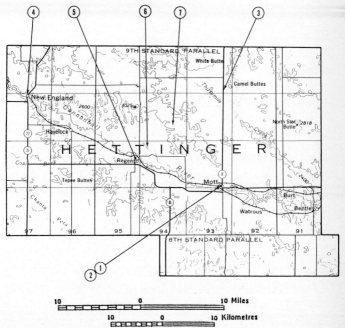

KIDDER COUNTY

1. Tappen area: east; in gravel pits.
   Agate, fortification. (10)

2. Tappen area: east; in natural gravel beds.
   Agate, fortification. Chalcedony. Jasper. (10-39)

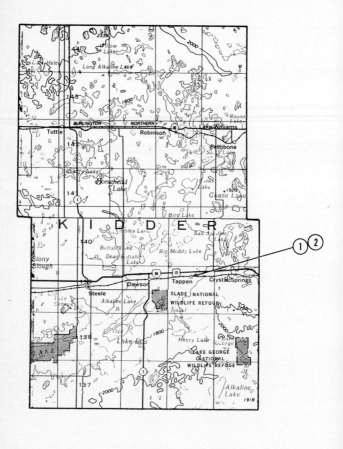

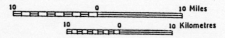

McHENRY COUNTY

1. Denbigh area: southwest along both sides of the Souris River all the way to Velva.

    Fossils, flora. Fossils, marine invertebrates. Quartz, rock: gem quality. Quartz, rose: gem quality. Quartz, smoky: gem quality. (3-7-18-20)

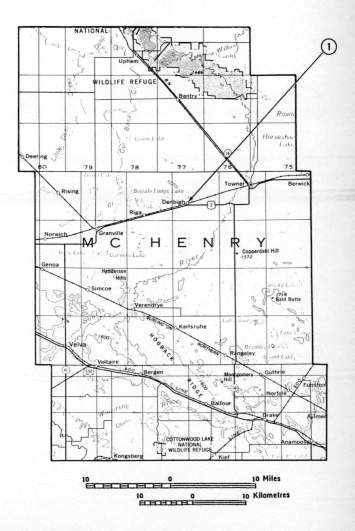

McKENZIE COUNTY

1. Cartwright area: just west on SR-200 to the Yellowstone River; collect upstream along the east shore of the Yellow-stone River to the Montana border.

    Agate, fortification. Agate, moss. (1-3-7-13-18-20)

2. Watford City area: 14 miles south on US-85 to the Little Missouri River; collect along both banks downstream to the Dunn County border and upstream along both banks above the southwest boundary of Theodore Roosevelt National Park, North Unit.

    Agate, fortification. Jasper. Petrified wood. Sard. (1-7-13-18-20)

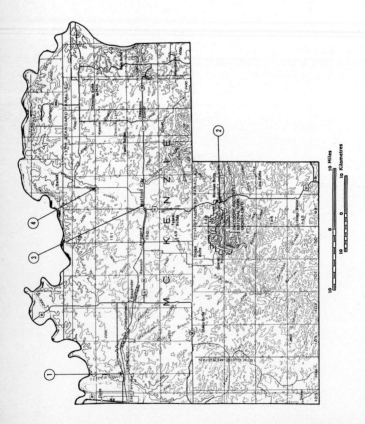

3. Watford City area: 0.75 mile east.

    Agate, moss. (10)

4. Watford area: 2.6 miles east on SR-73 to side road left; 9.1 miles north on latter to bridge over Tobacco Garden Creek; collect upstream along both sides of the creek to its head-waters.

    Petrified wood. (1-7-13)

MERCER COUNTY

1. Golden Valley area: just south; at the Crowley Flint Quarry.

    Flint: gemmy. (12)

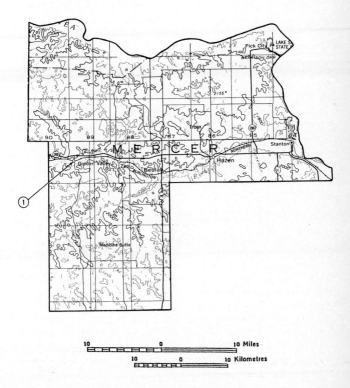

MORTON COUNTY

1. Mandan area gravel pits.

    Agate: nodules; small. Petrified wood: with teredo worm tracks. (7-10)

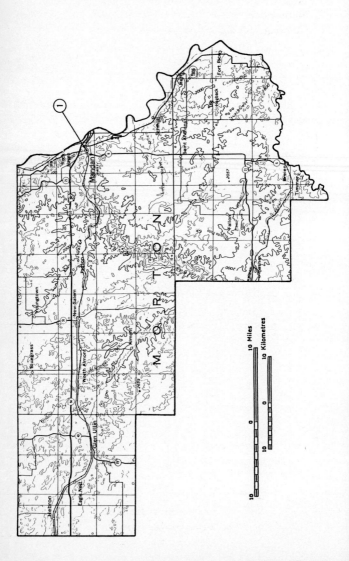

MOUNTRAIL COUNTY

1. Lostwood area: northwest throughout the area north of White Lake and westward to Powers Lake.

    Glauberite. Halite. (1-13-39)

2. Stanley area: just west in mining areas and along the shorelines of the saline lakes in this area.

    Glauberite. Halite. Thenardite. (2-20)

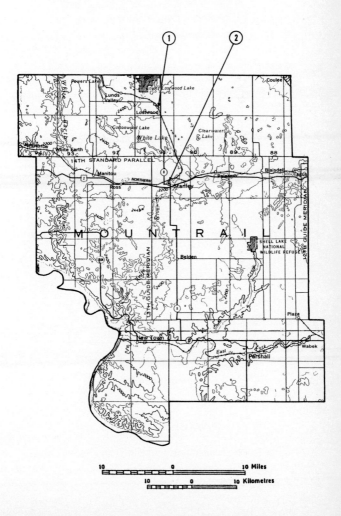

PEMBINA COUNTY

1. Concrete area abandoned limestone quarry.
   Agate, fortification. Chalcedony. Jasper. Petrified
   wood. (12)

2. Jollette area: 1.5 miles east; then upstream along the west
   bank of the Red River to the mouth of the Pembina River.
   Agate, fortification. (1-3-7-13-18-20)

3. Leroy area: throughout the Pembina Hills.
   Gypsum: good crystals. (1-13)

RAMSEY COUNTY

1. Devils Lake area: along shorelines of Devils Lake.
   Agate, fortification. Chalcedony. Jasper. Petrified
   wood. (3-20)

2. Devils Lake area: in dry-land gravel beds of surrounding area.

> Agate, fortification. Chalcedony. Jasper. Petrified wood. (1-3-13-39)

3. Devils Lake area: along shorelines of East Devils Lake.

> Agate, fortification. Chalcedony. Jasper. Petrified wood. (3-20)

4. Webster area: 2.1 miles west; along shorelines of Dry Lake.

> Agate, fortification. Chalcedony. Jasper. Petrified wood. (3-20)

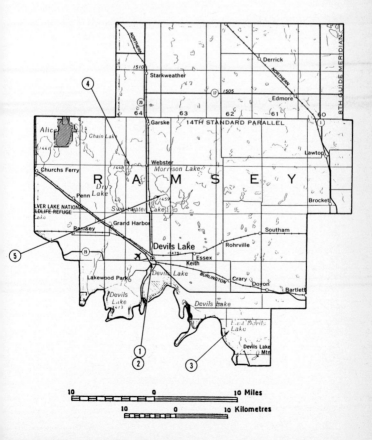

5. Webster area: 4 miles south-southeast; along the shore-lines of Sweetwater Lake.

> Agate, fortification. Chalcedony. Jasper. Petrified wood. (3-20)

RANSOM COUNTY

1. Anselm area.
   > Petrified wood. (1-13)

2. Lisbon area: upstream and down along both sides of the Sheyenne River.
   > Amethyst, fortification. Chalcedony. Jasper. Petrified wood: black; with teredo worm borings filled with deep yellow-orange chalcedony crystals. (1-3-13-18-20)

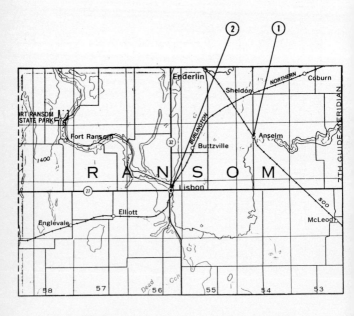

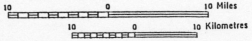

ROLETTE COUNTY

1. Dunseith area gravel pits.

 Agate, fortification. Chalcedony. Jasper. Petrified wood. (10)

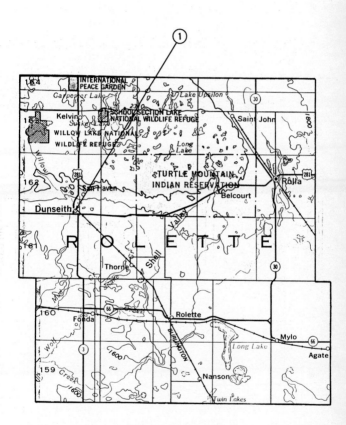

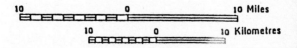

SLOPE COUNTY

1. Amidon area: 7.9 miles east and north on US-85 to right turn onto SR-21; 2.75 miles east on SR-21 to the Cannonball River; collect downstream (north and east) on both sides to the Hettinger County border and upstream to the headwaters.

    Jasper. Petrified wood. (1-3-7-13-20)

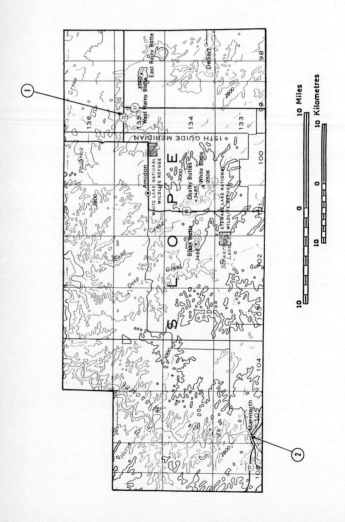

2. Marmarth area: northward along both sides of the Little Missouri River.

> Agate, fortification. Petrified wood. (1-3-7-13-18-20)

STARK COUNTY

1. Dickinson area.

> Chalcedony: pebbles. (1-3-26)

2. Dickinson area: 6 miles north on SR-22; on both sides of the road at the Dunn County border.

> Agate, fortification. Jasper. Petrified wood. (1-13)

3. Richardton area: collect for 8.5 miles south on both sides of SR-8 to the Heart (aka Green) River.

> Chalcedony. Jasper. Petrified wood. (1-13)

4. Richardton area: 8.5 miles south on SR-8 to the Heart (aka Green) River; upstream and down along both sides of the river.

> Agate, fortification. Chalcedony. Jasper. Petrified wood. (1-7-13-20)

5. Richardton area: 8.5 miles south on SR-8 to the Heart (aka Green) River; collect south along both sides of SR-8 to the Hettinger County border.

> Chalcedony. (1)

WARD COUNTY

1. Minot area: in gravel pits just south.

> Agate, fortification. Chalcedony. Jasper. Petrified wood. (10)

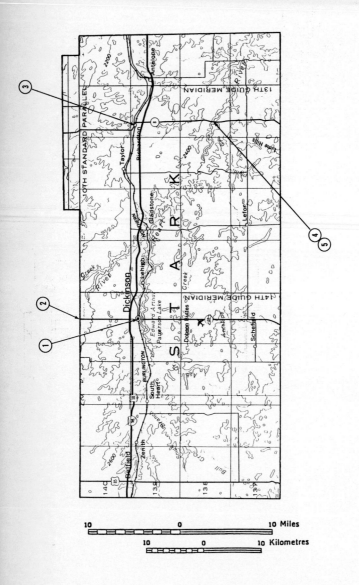

10 Miles

10 Kilometres

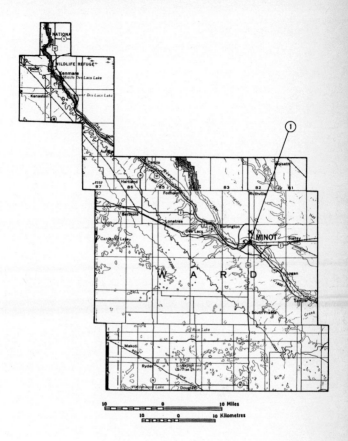

WILLIAMS COUNTY

1. Appam area: east-southeast along both sides of the Little
   Muddy River to its headwaters.

   Agate, fortification. Petrified wood. Selenite: waterworn crystals. (1-7-13-20)

2. Williston area: upstream along the north bank of the Missouri River from the mouth of the Little Muddy River to the Montana border.

   Agate, moss: translucent. Chalcedony. (20)

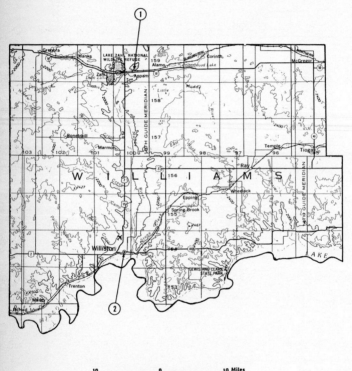

OREGON

Statewide listing of 190 locations in the following counties:

Baker (11)	Harney (11)	Morrow (1)
Benton (1)	Jackson (12)	Multnomah (1)
Clackamas (1)	Jefferson (16)	Polk (1)
Clatsop (3)	Josephine (2)	Sherman (4)
Columbia (3)	Klamath (2)	Tillamook (6)
Coos (4)	Lake (11)	Union (2)
Crook (22)	Lane (8)	Wallowa (2)
Curry (9)	Lincoln (5)	Wasco (9)
Deschutes (5)	Linn (8)	Wheeler (4)
Douglas (7)	Malheur (16)	
Grant (2)	Marion (1)	

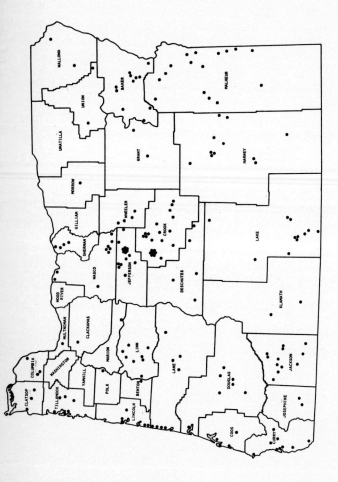

COLLECTING SITES IN OREGON

(FOR SPECIFIC DETAILS ON THESE
LOCATIONS, SEE THE FOLLOWING
COUNTY MAPS AND RELATED TEXT)

BAKER COUNTY

1. Baker area: in the Greenhorn district.

    Petrified wood: fernwood (*Tempskya* var.); in rolled masses or nodules to 50 lb; silicified; visible stems, roots, cell structure. (3-7-17)

2. Baker area: southeast; large triangular area from Baker to Richland to Durkee and back to Baker; abundant specimens.

    Agate. Chalcedony: geodes to 10″ diameter; quartz-lined. Jasper: dark green (aka Oregon jade or plasma). Petrified wood. (1-3-14-volcanic rock)

3. Durkee area.

    Opal, wood: fine. (3)

4. Durkee area: 2 miles north; in commercial gypsum quarry.

    Gypsum (aka Satin spar): fine. (12)

BAKER COUNTY
OREGON

DIVIDED INTO TWO
SEPARATE MAPS ON
THE FOLLOWING TWO
PAGES -- EACH MAP
IDENTIFIED BY ITS
NUMBER, AS SHOWN
IN DIAGRAM

5. Durkee area: 2.2 miles southeast to the mouth of Shirttail Creek; collect up that creek to its headwaters.

   Chalcedony. (7-18-20)

6. Durkee area: 2.2 miles southeast to Manning Creek; 5.75 miles upstream in Manning Creek to the mouth of Sutton Creek; collect upstream in Sutton Creek to its headwaters.

   Opal, wood: fine. (3)

7. Huntington area: near MM-393 on SR-30.

   Chalcedony: geodes to 9″ diameter; quartz-lined. (1-13)

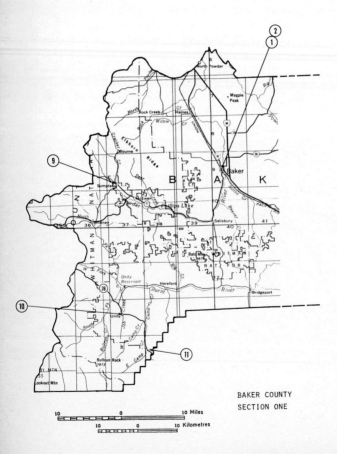

BAKER COUNTY
SECTION ONE

8. Pleasant Valley area.

   Garnet, almandine. Opal, wood: fine; banded black and white. (6-7)

9. Sumpter area: in Powder River and in adjacent dredging piles.

   Agate. (7-18-20-30)

10. Unity area.

    Agate. Chalcedony. Petrified wood. (1-13)

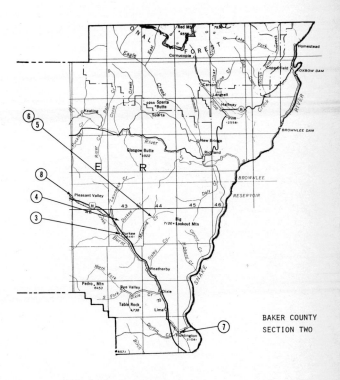

BAKER COUNTY
SECTION TWO

11. Unity area: 7.5 miles southeast on SR-26; just northeast of Eldorado Pass.

> Jasper. (1-13-19-rhyolite -29)

BENTON COUNTY

1. Greenberry area: 3.5 miles east; on the west bank of the Willamette River.

> Agate. Jasper. (7-20)

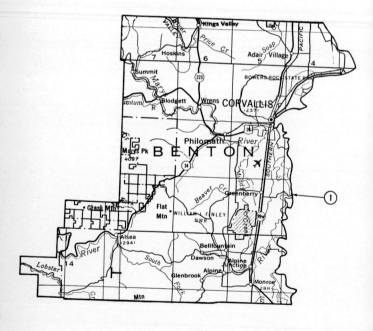

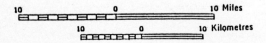

CLACKAMAS COUNTY

1. Oregon City area: upstream on both sides of the Clackamas River from its mouth.

    Jasper: green (aka Oregon jade or plasma). (3-7-18-20)

CLATSOP COUNTY

1. County-wide along coastline; especially at and near mouths of streams entering ocean.

    Agate. Jasper. (3-7-20)

2. Seaside area: along beaches.

    Agate. Jasper. (3-20-26)

3. Vinemaple area: in the Nehalem River.

    Carnelian: fine. (3-7-20)

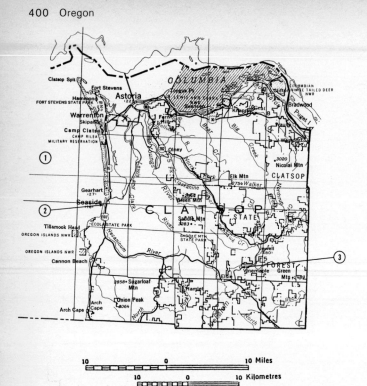

COLUMBIA COUNTY

1. Goble area.
   Chabazite. (5-basalt)

2. Vernonia area: at the Branch property.
   Agate, plume. Carnelian. (1-3-13-20)

3. Vernonia area: upstream along both sides of the Nehalem
   River to the mouth of Clear Creek.
   Agate. Carnelian. Chalcedony. Jasper. (7)

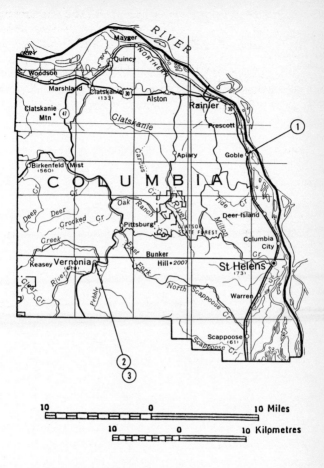

COOS COUNTY

1. Bullards area: at Bullards Beach State Beach.
   Agate. Petrified wood. (20-26)

2. Charleston area beaches, especially at Fossil Point.
   Agate: waterworn. Fossils, marine: fish bones; mammal bones; shark teeth; shells. Petrified wood: waterworn. (20)

3. Coos Bay area: south to Seven Devils Wayside Park; collect at Whiskey Run Beach and southward from there.

 Agate. Petrified wood: brown. (3-20-26)

4. County-wide along coastline and in beds of all streams entering the ocean.

 Agate. Jasper. (3-7-20-26)

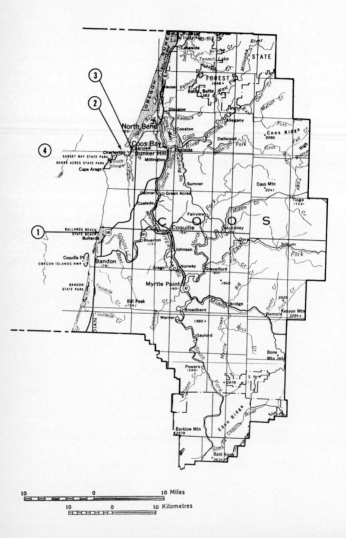

CROOK COUNTY

1. Post area: 1 mile east; cross bridge and turn left on Shotgun Road to Booton Ranch (CF).

   Agate (Thundereggs). Petrified wood. (1)

2. Post area: 6 miles northwest; at Bonnieview Ranch (CF).

   Petrified wood. (1)

3. Post area: 8 miles east to MM-33; turn right across wooden bridge; continue south on dirt road across five cattle guards to end, where there is a sign for national forest and another sign pointing west to diggings; turn right and go west to first road right; turn there and go 1 mile north to another right turn; turn there and go 1 more mile to Maury Mountain diggings.

   Agate (Thundereggs): green, red. Agate, moss. (1-13-39)

4. Post area: 14 miles east; at the Dick Ranch (CF).

   Petrified wood. (1)

5. Post area: 18 miles east to Camp Creek Road; turn south; then west to collect on plateau in area between creeks and on both sides of the South Fork of the Crooked River.

   Petrified wood: branches; dendritic. (1-7-13-20)

6. Powell Butte area: at mine on steep hillside of Carey Ranch (CF).

   Agate, plume (aka Flame agate or Carey plume agate); pieces to 4 lb. (2)

7. Prineville area: east; east of Wildcat Mountain, on Stevens Mountain. (CF in some areas.)

   Agate (Thundereggs): to 6″ diameter. (1-13)

8. Prineville area: east; on McAllister Butte near the Ochoco River.

   Agate, moss: fine; brown. (3-39)

9. Prineville area: east; on Wildcat Mountain (CF).

   Agate (Thundereggs): to 5″ diameter. (1-13-39)

10. Prineville area: all along both sides of US-26, northwest to the Jefferson County border.

>   Agate (Thundereggs). (1-7-13-20)

11. Prineville area: on both sides of SR-126, southwest to the Deschutes County border.

>   Agate (Thundereggs). Chalcedony. Jasper. Petrified wood. (1-3-13-39)

12. Prineville area: southeast on Juniper Canyon Road; at northwest end of reservoir; go west on road to its end and collect west.

>   Agate, moss. (1-13-20)

13. Prineville area: 5 miles east on US-26; along shoreline of Ochoco Lake (especially directly across the lake from the boat landing).

>   Jasper. (20)

14. Prineville area: 9 miles east on US-26 to Mill Creek Road (at store); turn left and go north 10 miles; turn right and cross bridge, then 4.5 miles more to road junction; stop here and walk straight ahead 0.2 mile; begin collecting at White Fir Spring and extend operations outward from there.

>   Agate (Thundereggs): to 5″ diameter. (1-13-39)

15. Prineville area: 11 miles southeast; at two pits, each 100′ in diameter; collecting only through arrangement with owners.

>   Agate, plume (aka Crooked Creek dendritic agate): small; dendritic with red or black plumes in clear chalcedony. (2-5-volcanic rock -19)

16. Prineville area: 16.5 miles east on US-26; turn off at angle onto gravel road and go 3 miles to Wolf Creek Road; take it to Sheep Creek sign and follow Cadle Road to Arvid Nelson Road; turn left and follow to the 21-mile sign; collect at Sheep Creek.

>   Agate, white. (1-3-7-20)

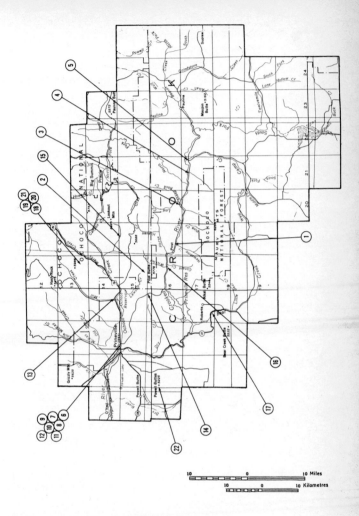

17. Prineville area: 18 miles east on Combs Flat Road (Post Road) to Eagle Rock; go right past rocks and 1.25 miles up sharp incline.

Agate, angel-wing. Agate: clear, dendritic. (1)

18. Prineville area: 19 miles south; 2 miles south of junction of Bear Creek and Crooked Creek; west of Bear Creek, on east slope of Taylor Butte.

> Agate: nodules to 3″ diameter; fine; dark green; inclusions are like suspended, filmy moss; some with quartz-lined cavities. (1-3)

19. Prineville area: 31 miles east on US-26 to the Marks Creek Guard Station; turn left on CR-27 and go 1 mile north; then northwest of CR-123 to CR-1203; follow the latter north to CR-1256 and turn right; go east to the campground; collect on the slopes of Ochoco Mountain.

> Agate, moss: clear, with dark green dendritic inclusions shaped like tiny fir trees. (1-7-13-20)

20. Prineville area: 31 miles east on US-26 to the Marks Creek Guard Station; turn left and follow Viewpoint signs on CR-125 to CR-123 (at sign); thence to the Lucky Strike claim.

> Agate (Thundereggs): to 5″ diameter. (1-13-39)

21. Prineville area: 31 miles east on US-26 to the Marks Creek Guard Station; follow Viewpoint signs on CR-127 to Whistler Spring.

> Agate, moss. (1-13)

22. Prineville area: 31 miles east on US-26 to the Marks Creek Guard Station; follow Viewpoint signs on CR-125 to CR-123 (at sign); thence to the Valley View claim.

> Agate (Thundereggs): to 5″ diameter. (1-13)

CURRY COUNTY

1. Agness area: in Rogue River and its canyon.

> Agate. Carnelian. Chalcedony. Garnet, grossularite. (3-7-14-20)

2. Cape Blanco area.

> Platinum: nuggets smaller than peas. (17)

3. County-wide along coastline, especially at mouths of all streams.

   Agate. Jasper. (7-18-20-26)

4. Gold Beach area: at the mouth of Rogue River and on beaches and in coves north to Nesika Beach.

   Agate. Garnet, grossularite: massive. Jasper. Petrified wood. (7-18-20-26)

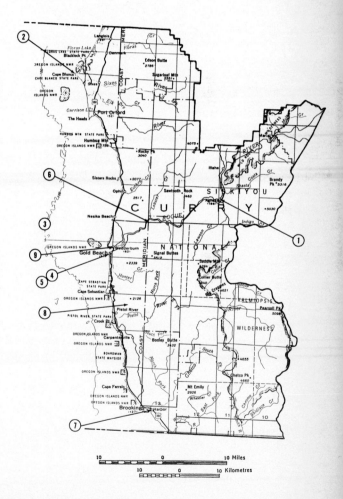

5. Gold Beach area: south on SR-101 to Lone Ranch State Park; at Rainbow Cove; around the point and at the southern terminus of the beach.

   Agate. (3-20-26)

6. Gold Beach area: 7 miles northeast on Rogue River Road to the Huntley Park area.

   Petrified wood. (18-20)

7. Harbor area: at the mouth of the Chetco River and upstream from there.

   Agate. Idocrase (Vesuvianite/Californite). Jasper. (7-20)

8. Pistol River area: take Forest Service Road to base of Sugarloaf Peak; collect there and in streams draining Sugarloaf Mountain.

   Jade, nephrite. (7-8-20)

9. Wedderburn area: collect in and along Rogue River for 3 miles upstream.

   Garnet, grossularite (aka Oregon jade): from pebbles to boulders of 5 lb. (3-7-20)

DESCHUTES COUNTY

1. Bend area: southeast on US-20 to MM-79; turn right on old road to Glass Buttes.

   Obsidian: many colors; fair quality (on surface) to highest quality (needing to be removed from matrix). (1)

2. Brothers area: on Sugarloaf Mountain.

   Quartz, rock. (1-13)

3. Hampton area: 4.5 miles northwest on US-20 to road at right; follow latter north to Hampton Butte; collect on unnamed butte directly west.

   Agate, moss: green. Jasper. Petrified wood. (1)

DESCHUTES COUNTY
OREGON

DIVIDED INTO TWO
SEPARATE MAPS
-- EACH MAP
IDENTIFIED BY ITS
NUMBER, AS SHOWN
IN DIAGRAM TO LEFT

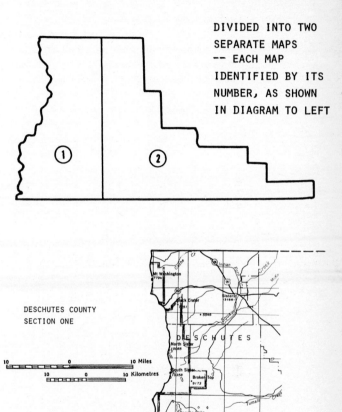

DESCHUTES COUNTY
SECTION ONE

4. La Pine area: 8.6 miles west on service road to the Wickiup Reservoir Dam; collect below dam and downstream in the Deschutes River.

   Agate: black. (7-20-28)

5. La Pine area: 5.8 miles north-northeast on US-97; then 12 miles east on the road leading to Paulina Lake; collect on both sides of the road between Paulina Lake and East Lake.

   Obsidian: most varieties. (1)

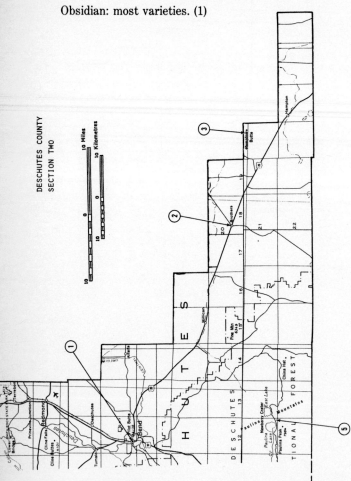

DOUGLAS COUNTY

1. County-wide along ocean beaches; especially at and near the mouths of streams.

   Agate. Jasper. Petrified wood. (7-20-26)

2. Idleyld Park area: on the North Umpqua River.

   Petrified wood: fine; with teredo worm borings (including some fossilized worms) lined with chalcedony; up to 500 lb. (7-20)

3. Melrose area: upstream from the confluence along both sides of both branches of the Umpqua River.

   Agate. Carnelian. Chalcedony. Jasper. (7-20)

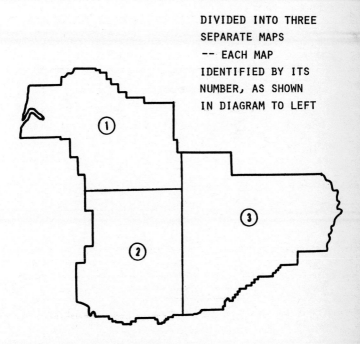

DOUGLAS COUNTY
OREGON

DIVIDED INTO THREE
SEPARATE MAPS
-- EACH MAP
IDENTIFIED BY ITS
NUMBER, AS SHOWN
IN DIAGRAM TO LEFT

412 Oregon

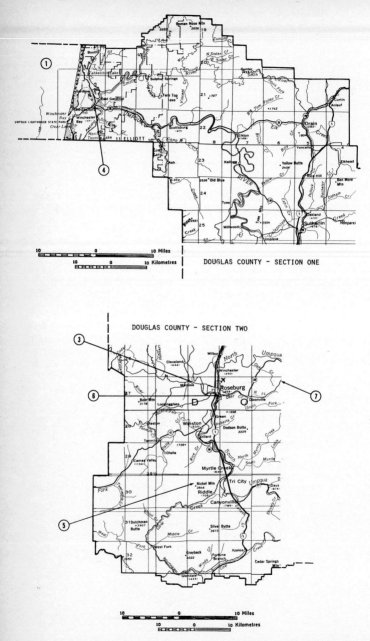

DOUGLAS COUNTY - SECTION ONE

DOUGLAS COUNTY - SECTION TWO

4. Reedsport area: upstream along the Umpqua River.

> Garnet, grossularite (aka Oregon jade); from pebbles to boulders of 10 lb. (3-7-18-20)

5. Riddle area: 4 miles northwest; on slopes of Nickel Mountain; at the nickel mines.

> Chrysoprase: bright green. (2-5-massive nickel -19)

6. Roseburg area: upstream along both sides of the Umpqua River to Glide.

> Agate. Petrified wood. (7-18-20)

7. Roseburg area: 12 miles east; in tributary of Deer Creek.

> Jasper: orbicular: fine; opalescent orbs in translucent gray fibrous quartz; usually round. (1-3-7-13-20)

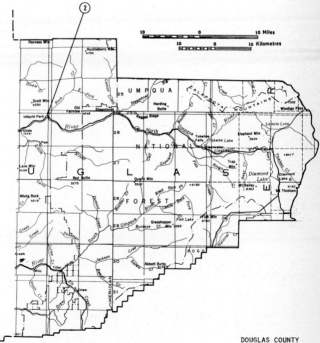

DOUGLAS COUNTY
SECTION THREE

GRANT COUNTY

1. Austin area: downstream on both sides of the Middle Fork
   of the John Day River, especially in all tributaries entering
   this river between Austin and Galena.

   Petrified fern. (3-7-18-20)

2. Canyon City area: upstream (south) along both sides of
   Canyon Creek.

   Serpentine. (3-7-18-20)

GRANT COUNTY
OREGON

DIVIDED INTO TWO
SEPARATE MAPS ON
THE FOLLOWING TWO
PAGES -- EACH MAP
IDENTIFIED BY ITS
NUMBER, AS SHOWN
IN DIAGRAM TO LEFT

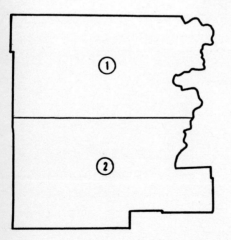

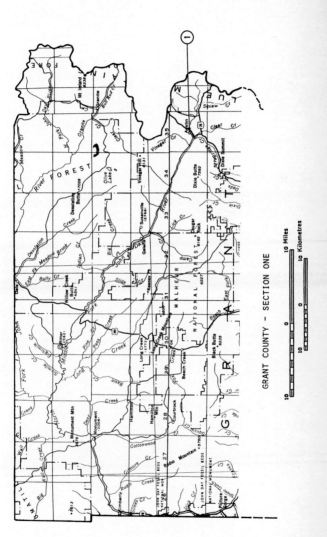

GRANT COUNTY – SECTION ONE

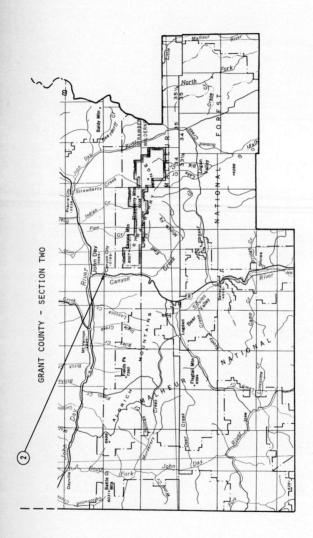

GRANT COUNTY – SECTION TWO

10 Miles

10 Kilometres

HARNEY COUNTY

1. Andrews area: in the Alvord Desert; southeast side of Steens Mountains; just west of Alvord Ranch; on the road between Dwenio and Folly Farm.

    Agate (Thundereggs): to 6″ diameter. (1-13)

2. Buchanan area: 4 miles south; at Robbins Thunderegg Diggings (CF).

    Agate (Thundereggs): to 5″ diameter. (1-13)

3. Burns area: at MM-167; on north side of road at base of hill (CF).

    Agate, plume. (1)

4. Burns area: east on SR-78 to SR-205; south on SR-205 to dirt road right just beyond MM-16; follow dirt road west for 2 miles; turn north for 2 miles to Wright Ranch.

    Agate, snakeskin. (1)

5. Burns area: west several miles.

    Agate (Thundereggs): to 6″ diameter. (1-13)

6. Burns area: 18 miles north-northeast on US-395 to rough road left; follow it 7 miles west to Myrtle Park; in Sylvies Canyon.

    Opal, wood. (1-13)

7. Drewsey area: 2 miles south to right turn on US-20; 2.25 miles southwest on US-20 to side road at left that leads to Warm Springs Reservoir; 10.6 miles south-southeast to first rough road at right; turn right and go past two washes to the first hilltop beyond the second wash.

    Agate, plume. (1-13)

8. Drewsey area: 10 miles south-southeast; in broad area covering the Harney-Malheur county line near the Warm Springs Reservoir.

    Agate (Thundereggs): to 6″ diameter. (1-13)

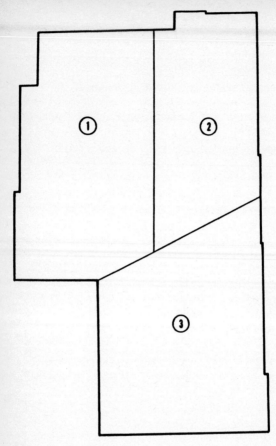

HARNEY COUNTY
OREGON

DIVIDED INTO THREE
SEPARATE MAPS ON
THE FOLLOWING THREE
PAGES -- EACH MAP
IDENTIFIED BY ITS
NUMBER, AS SHOWN
IN DIAGRAM

HARNEY COUNTY
SECTION ONE

10   0   10 Miles

10   0   10 Kilometres

9. Stinkingwater Pass area: from summit on US-20, turn south on BLM road and go 4 miles to rough road heading east; follow latter to diggings.

Petrified wood. (1-13)

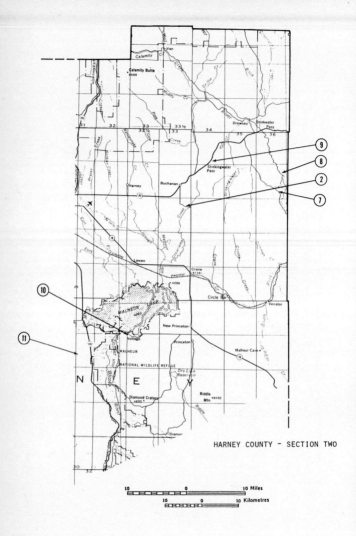

HARNEY COUNTY - SECTION TWO

10. Voltage area: collect on both sides of road for 2.4 miles
west to the north end of the Malheur Wildlife Refuge.
Agate, snakeskin. (5-pumice)

HARNEY COUNTY - SECTION THREE

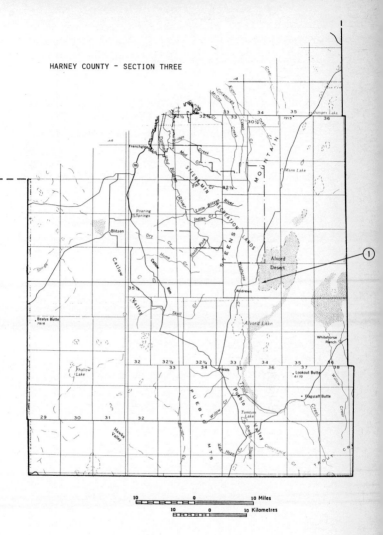

11. Voltage area: 7 miles west to SR-205; south on SR-205 to
     MM-32; east of the road, at Saddle Butte.
         Petrified wood. (1-13)

JACKSON COUNTY

1. Ashland area: east; on slopes of Greensprings Mountain.
   Agate (Thundereggs): to 5″ diameter. Carnelian. (1-13)

2. Ashland area: southeast to junction of I-5 and SR-66; then 7.5 miles east on SR-66; in road cut and adjacent hills.
   Agate (Thundereggs): to 5″ diameter; green. (1-4-13)

3. Butte Falls area: from SR-234, go south on Gunderson Road to the RR tracks; walk west and collect in ravine.
   Jasper: green. (1-13)

4. Butte Falls area: northwest to broad area of collecting bed centered just southwest of Crow Foot.
   Agate (Thundereggs): to 7″ diameter. Agate, moss. Jasper. (1-7-13-20-39)

5. Brownsboro area: collect for 8 miles northeast on both sides of road, especially in ravines, washes, and creeks southwest.
   Agate. Petrified wood. (1-3-7-13-20)

6. Eagle Point area: collect upstream along both sides of the Little Butte River for 8 miles.
   Agate, moss: very fine. (3-7-18-20)

7. McLeod area.
   Natrolite: compact masses. (14-basalt)

8. McLeod area: at Big Butte.
   Agate (Thundereggs): to 5″ diameter. Jasper: green, white. (1-3-13)

9. Pinehurst area: upstream and down along both sides of Jenny Creek.
   Agate. (3-7-18-20)

10. Table Rock area: collect on the slopes of Table Rock Mountain, especially in all streams and run-off areas on the north slope.

   Petrified wood: fine quality. (1-7-13-20)

11. White City area: 2.4 miles northeast on SR-62 to Antelope Creek bridge; then collect upstream along both sides of creek for 10 miles.

   Agate: transparent; dendritic. (3-7-18-20)

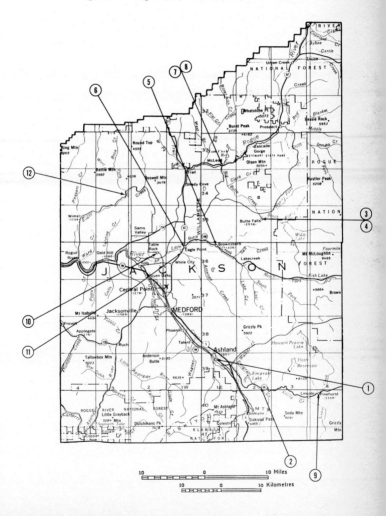

12. **Wimer area:** 8.5 miles east-northeast on the road along Evans Creek; in road cut and mine.

>   Rhodonite. (2-4-7-20)

JEFFERSON COUNTY

1. Ashwood area: at Bedortha Ranch (CF).
   Agate (Thundereggs). (1-13-39)

2. Ashwood area: at Brown Ranch (CF).
   Agate (Thundereggs). (1-13-39)

3. Ashwood area: at Friend Ranch (CF).
   Agate (Thundereggs). (1-13-39)

4. Ashwood area: at Indian Creek Ranch (CF).
   Agate (Thundereggs). (1-13-39)

5. Ashwood area: at Keegan (CF).
   Agate (Thundereggs). (1-13-29)

6. Ashwood area: at McDonald Ranch (CF).
   Agate (Thundereggs). (1-13-39)

7. Ashwood area: at Swanson Ranch (CF).
   Agate (Thundereggs). (1-13-39)

8. Ashwood area: 9 miles northeast; at Norton Ranch (CF).
   Agate (Thundereggs): purple, green, blue, pink. Petrified wood: chunks; limb casts. (1-13-39)

9. Kilts area: 9.7 miles east; in Cherry Creek and downstream to where it empties into the John Day River at the Wheeler County border.
   Agate (Thundereggs): to 6″ diameter. Chalcedony. Jasper. Petrified wood: fernwood. (1-3-7-13-20)

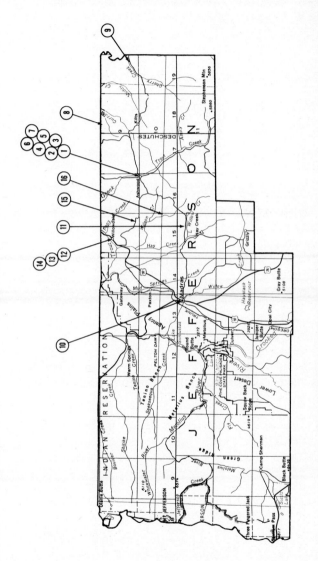

10. Madras area: large, roughly trapezoidal area of Thunderegg beds, bordered by Madras, Willowdale, Ashwood, and Hay Creek; excellent specimens also abundant a few miles northeast of Ashwood.

>   Agate (Thundereggs): to 6″ diameter. (1-3-7-13-20-39)

11. Madras area: 11 miles east; in Willow Creek and adjacent hills.

>   Agate (Thundereggs): to 5″ diameter. Jasper. (1-7-13-20)

12. Willowdale area: just south on US-97; at the Kennedy Ranch (CF).

>   Agate (Thundereggs): to 6″ diameter. Petrified wood. (1-13-39)

13. Willowdale area: just south on US-97; at the Richardson Ranch (CF).

>   Agate (Thundereggs): to 6″ diameter. Petrified wood. (1-13-39)

14. Willowdale area: 1 mile north on US-97, then 7 miles east on Antelope Road; at Folmsbee Ranch.

>   Agate (Thundereggs). Petrified wood. (1-13-39)

15. Willowdale area: 4 miles southeast; at Priday Ranch (CF).

>   Agate (Thundereggs): to 6″ diameter. Opal, precious; occasional. (1-13-39)

16. Willowdale area: 8 miles south-southeast; at Palmer Eagle Ranch (CF).

>   Agate (Thundereggs). Agate, moss. (1-13-39)

JOSEPHINE COUNTY

1. Cave Junction area: 11 miles east on SR-46; 3.2 miles before entrance to Oregon Caves National Monument; collect upstream along Sucker Creek to its headwaters.

>   Rhodonite: pebble-size to small boulders. (7-19-to 4″ thick -20)

2. Holland area: 0.8 mile south; upstream and down along Althouse Creek.

Agate. Garnet, grossularite (aka Oregon jade): boulders to 5 lb. Jasper. Serpentine. (3-7-20)

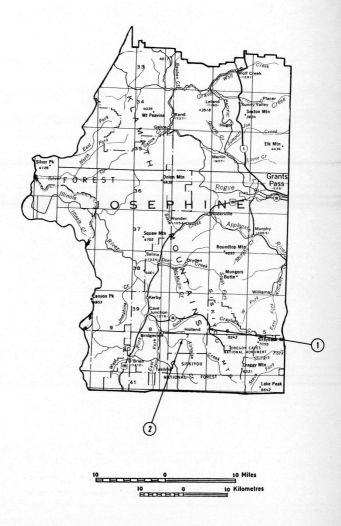

KLAMATH COUNTY

1. Fort Klamath area: 7 miles northwest on SR-62; 7 miles south of Crater Lake; just south of the south boundary of Crater Lake National Park.

    Agate. Jasper (aka Crater Lake flower jasper): fine. (3)

2. Klamath Falls area.

    Opal, green (aka Hyalite): fluorescent. (5-9-basalt)

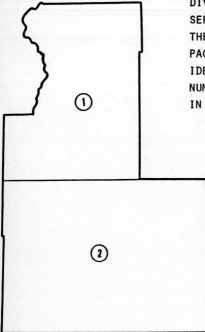

KLAMATH COUNTY
OREGON

DIVIDED INTO TWO
SEPARATE MAPS ON
THE FOLLOWING TWO
PAGES -- EACH MAP
IDENTIFIED BY ITS
NUMBER, AS SHOWN
IN DIAGRAM TO LEFT

KLAMATH COUNTY - SECTION ONE

10   0   10 Miles

10   0   10 Kilometres

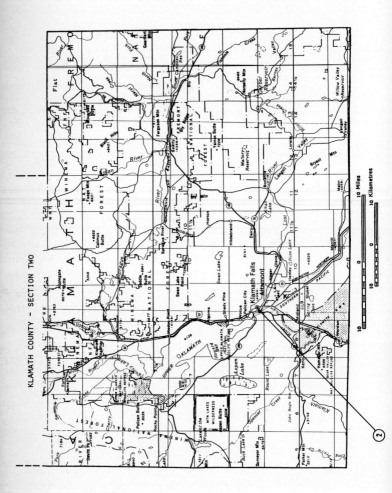

KLAMATH COUNTY – SECTION TWO

LAKE COUNTY

1. Lakeview area.
   Sanidine. Sanidine sunstone. (6)

2. Lakeview area: in both Bullard Canyon and Deadman Canyon.
   Agate (Thundereggs). Jasper: orbicular. (1-13)

3. Lakeview area: west; in Dry Creek.
   Agate (Thundereggs). (1-7-13)

4. Lakeview area: 4 miles south on US-395; then 1 mile east in hills.
   Agate (Thundereggs): blue. (1-13)

LAKE COUNTY
OREGON

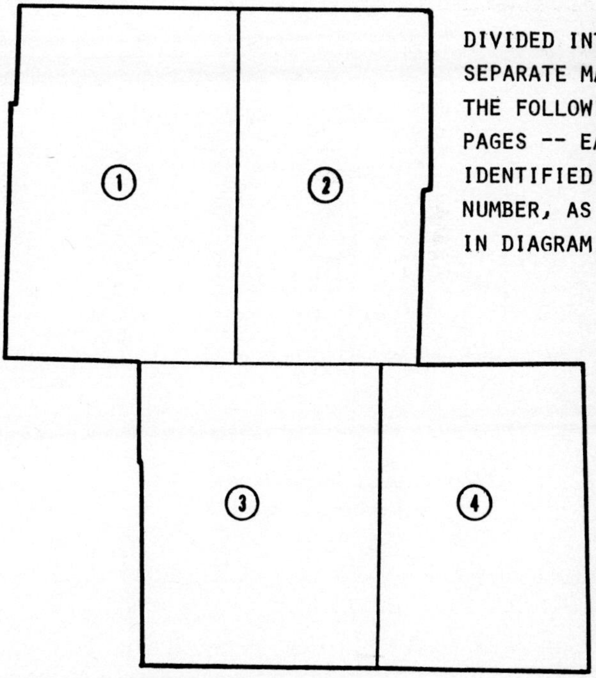

DIVIDED INTO FOUR
SEPARATE MAPS ON
THE FOLLOWING FOUR
PAGES -- EACH MAP
IDENTIFIED BY ITS
NUMBER, AS SHOWN
IN DIAGRAM TO LEFT

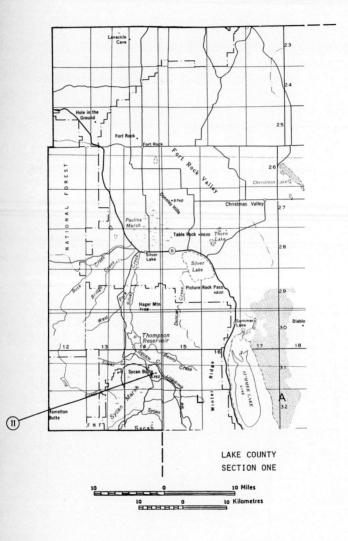

LAKE COUNTY
SECTION ONE

10   0   10 Miles

10   0   10 Kilometres

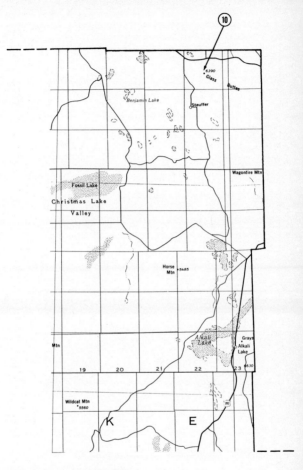

LAKE COUNTY - SECTION TWO

LAKE COUNTY - SECTION THREE

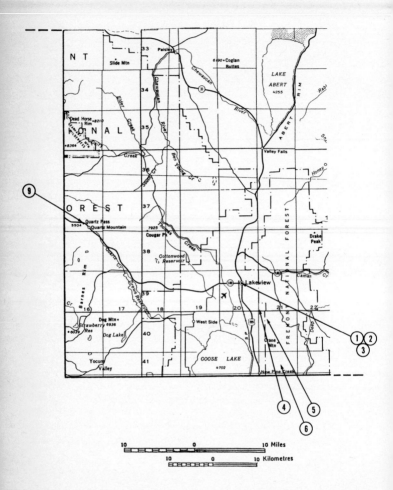

5. Lakeview area: 5 miles south on US-395; then east to Crane Canyon.

   Agate (Thundereggs): to 5″ diameter. (1-13)

6. New Pine Creek area: 3.2 miles east-northeast; in south Crane Canyon and the surrounding general area, southward into California.

   Agate (Thundereggs): to 5″ diameter. (1-13)

LAKE COUNTY -   SECTION FOUR

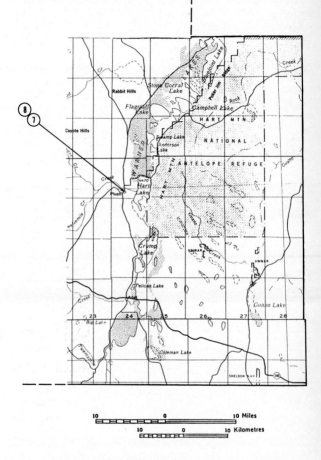

```
          10           0              10 Miles
          10           0              10 Kilometres
```

7. Plush area: 0.8 mile north to fork; bear right toward Hart
   Mountains; from fork go another 18 miles (northeast) to
   Hart Mountain National Wildlife Refuge headquarters;
   turn south to Hart Mountain; walk to top (accessible by
   road by difficult route on east slope) and collect on rims and
   slopes of five west-facing canyons.

   Agate: geodes to 8″ diameter. Agate (Thundereggs):
   to 6″ diameter. Opal, precious: fire. (1-5-basalt -6-13-
   14)

8. Plush area: 0.8 mile north to fork; bear right on road that goes toward Hart Mountains; take that right fork only 0.5 mile and then take a left fork for another 5 miles; then walk 0.5 mile up dry wash to Rabbit Basin in the Warner Valley.

   Labradorite, aventurine (aka Sunstone). (1-5-basalt -14)

9. Quartz Pass area.

   Agate (Thundereggs). Petrified wood. (1-3-4-13)

10. Stauffer area: 7 miles north to right turn on US-20; then 2.8 miles east on US-20; 12 miles southeast of Hampton (Deschutes County); 2 miles due south of the Deschutes County border; along US-20 in the area of Glass Buttes.

    Obsidian: massive; black, red, pink, purple, copper, blue, gold, green; solid-colored, banded, spiderweb, swirled, rainbow; very abundant. (1-13)

11. Sycan Butte area.

    Agate (Thundereggs). Jasper. Petrified wood. (1-13)

LANE COUNTY

1. County-wide along the coastline, especially at the mouths of all streams entering the ocean.

   Agate. Jasper. (3-7-20)

2. Finn Rock area: from its mouth at the McKenzie River, collect upstream (south) on both sides of Quartz Creek to its headwaters.

   Agate. Jasper. (1-3-7-13-20)

3. Florence area: 11 miles north on US-101; 9 miles south of the Lincoln County border; at Heceta Head in Devil's Elbow State Park; in and near creek mouth.

   Agate. Jasper. Petrified wood. (3-7-20-26)

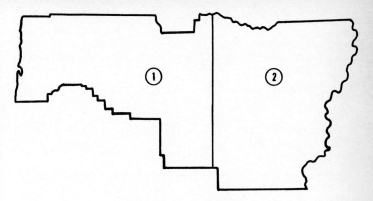

LANE COUNTY, OREGON

DIVIDED INTO TWO SEPARATE
MAPS ON THE FOLLOWING TWO
PAGES -- EACH MAP IDENTI-
FIED BY ITS NUMBER, AS
SHOWN IN DIAGRAM ABOVE

4. Florence area: 13.75 miles north on US-101; 6.25 miles
   south of the Lincoln County border; in and near the mouth
   of Big Creek.
   > Chalcedony: geodes to 8″ diameter; some enhydros;
   > clear; sagenitic inclusions. Fossils, marine: corals; si-
   > licified. Garnet, grossularite (aka Oregon jade); from
   > pebbles to boulders of 5 lb. Jasper: orbicular. Petrified
   > wood. (1-3-7-13-20)

5. Florence area: 16.8 miles north on US-101; 3.25 miles south
   of the Lincoln County border; in and near the mouth of
   Tenmile Creek.
   > Chalcedony: geodes to 8″ diameter; some enhydros;
   > clear; sagenitic inclusions. Fossils, marine: corals; si-
   > licified. Jasper: orbicular. Petrified wood. (1-3-7-13-20)

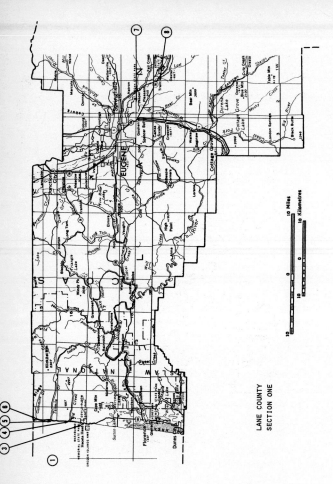

LANE COUNTY
SECTION ONE

6. Florence area: 19.6 miles north on US-101; 0.75 mile south
of the Lincoln County border; in and near the mouth of
Cummins Creek.

Chalcedony: geodes to 8″ diameter; some enhydros;
clear; some with sagenitic inclusions. Fossils, marine:
corals; silicified. Jasper: orbicular. Petrified wood.
(1-3-7-13-20)

LANE COUNTY
SECTION TWO

7. Goshen area: 3 miles east; on Mount Pisgah.

Agate. Calcite. Heulandite. Mesolite: silky; massive; chatoyant; abundant. Quartz, rock. (1-9-volcanic rocks -13-29)

8. Trent area: southeast on SR-58; at Snyder Ranch (CF).

Agate. Realgar. (1-13)

LINCOLN COUNTY

1. County-wide along the coastline, especially near the mouths of streams.

    Agate. Jasper. Petrified wood. (3-7-20)

2. Newport area: at Beverly Beach.

    Agate. Jasper: some orbicular. Petrified wood. (3-7-20)

3. Newport area: north.

    Petrified wood: logs with worm borings; gray; water-worn. (7-20)

4. Yachats area: at the mouth of the Yachats River.

    Chalcedony: geodes to 8″ diameter; some enhydros; clear; some with sagenitic inclusions. Fossils, marine: corals; silicified. Jasper: some orbicular. Petrified wood. (3-7-20)

5. Yachats area: 2 miles north of the mouth of China Creek.

    Chalcedony: geodes to 8″ diameter; some enhydros; clear; some with sagenitic inclusions. Fossils, marine: corals; silicified. Jasper: orbicular. (20)

LINN COUNTY

1. Lebanon area: south on US-20 to left turn at Tin Road; follow it to next left; make turn and then take next right onto Tye Road; follow Tye Road to the Dolly Drummond Agate Beds (CF).

    Agate. Carnelian. Jasper. (1-6-13)

2. Peoria area: on midstream gravel bars and along the east shore of the Willamette River downstream to Independence (in Polk County).

    Agate. Carnelian. Jasper. (3-7-18-20)

3. Scio area: 2 miles south on SR-226 to right turn at school-house; go 1 mile west to right turn at dirt lane; go 1 mile north to farmhouse (CF).

    Jasper. Petrified wood. (1-3-13-39)

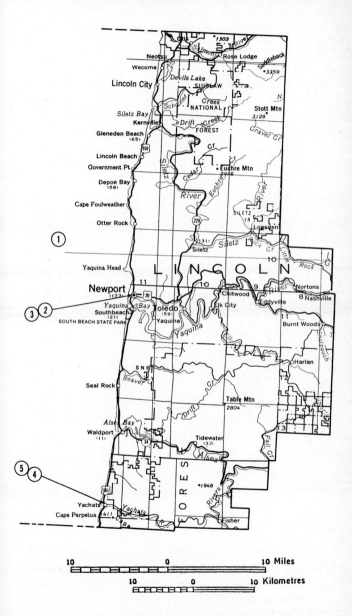

Otis
•1303
Rose Lodge
Saddleback
•3359
Neotsu
Wecoma
Devils Lake
Lincoln City
SIUSLAW
Siletz Bay
Creek
NATIONAL
Stott Mtn
3128•
Kernville
Drift
Creek
Gleneden Beach
(65)
FOREST
Gravel Cr
101
Lincoln Beach
Cr
Euchre Mtn
Government Pt
Cedar
2446
Depoe Bay
(58)
River
Euchre
Cape Foulweather
River
SILETZ
Otter Rock
229
I R
Logsden
131
Siletz
Siletz L
Rock
Cr
Little
Yaquina Head
L I N C O L N
10
Rock
Newport
(172)
Chitwood
9
River
Nortons
20
11
Elk City
Eddyville
8 Nashville
Yaquina
Bay
Toledo
Southbeach
(59)
11
Burnt Woods
(21)
Yaquina
SOUTH BEACH STATE PARK
Yaquina
Harlan
S N B
Seal Rock
Beaver
Table Mtn
Drift
2804
13
Alsea Bay
Waldport
34
Tidewater
(11)
F O R E S
(32)
Alsea
River
5 4
•1946
Yachats
Rivers
Cape Perpetua
(411)
Yachats
Fisher
Cr

10   0   10 Miles

10   0   10 Kilometres

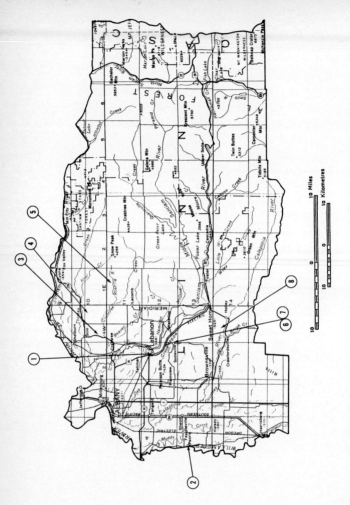

4. Scio area: 3 miles east on SR-226 to right turn on Richardson Gap Road; follow this to next paved road on left; follow
   paved road east 1.5 mile to first gravel road on left; take
   gravel road to end; at Rogers Mountain.

   Petrified wood. (1-7-13-20)

5. Scio area: 9 miles southeast; at the Prospect Mountain Mine (CF).

    Agate. Petrified wood. (1-2-13)

6. Sodaville area: at the Moore Ranch (CF).

    Carnelian: good quality. (1)

7. Sodaville area: at the Tyler Ranch (CF).

    Carnelian: good quality. (1)

8. Sweet Home area: 4 miles southwest on SR-228 to Holley School, at which turn right and go just over 2 miles to farmhouse; from there go on foot along logging road for 1 mile.

    Agate, blue. (1)

MALHEUR COUNTY

1. Adrian area: south; in diggings along irrigation canal and west of there at Alkalin Lake.

    Agate (Thundereggs): to 6″ diameter. Petrified wood: red. (1-13)

2. Brogan area.

    Agate (Thundereggs): to 5″ diameter. Chalcedony. Petrified wood. (1-13)

3. Crowley area: 6.9 miles northeast; in the Skull Springs area.

    Agate: geodes to 6″ diameter. Agate (Thundereggs): to 5″ diameter. Opal, wood. (1-13)

4. Crowley area: 11 miles northeast to dry bed of Butte Creek; then 0.5 mile west; collect in creekbed and adjacent areas.

    Agate (Thundereggs): to 5″ diameter. (1-13)

5. Crowley area: 17.4 miles northeast to second crossing of Dry Creek; collect upstream (west) for 5 miles.

    Agate (Thundereggs): to 5″ diameter. (1-13)

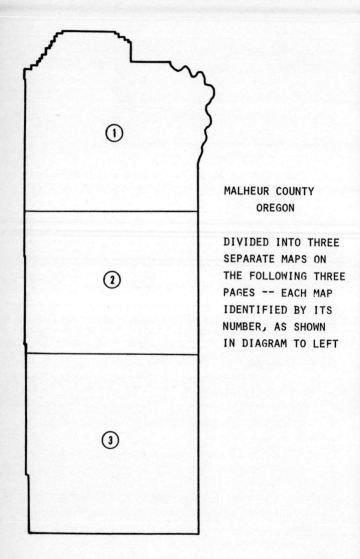

MALHEUR COUNTY
OREGON

DIVIDED INTO THREE
SEPARATE MAPS ON
THE FOLLOWING THREE
PAGES -- EACH MAP
IDENTIFIED BY ITS
NUMBER, AS SHOWN
IN DIAGRAM TO LEFT

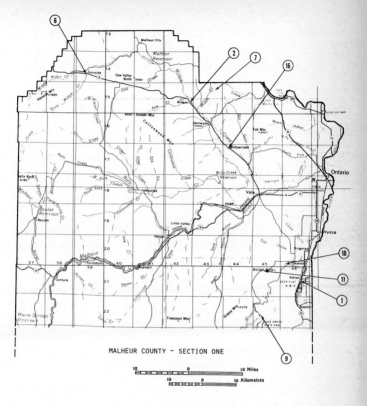

MALHEUR COUNTY - SECTION ONE

6. Ironside area.

   Agate (Thundereggs): to 5″ diameter. Chalcedony.
   Petrified wood. (1-13)

7. Jamieson area: northeast to sign for Huntington; then 6
   miles north.

   Petrified wood (aka "Bogwood Agate"). (1-13)

8. Jordan Valley area.

   Chert: very colorful. (1-3)

9. Owyhee Dam area: below dam and downstream (north and
   then northeast) on both sides.

   Agate. Opal, wood. Petrified wood. (1-3-7-13-20)

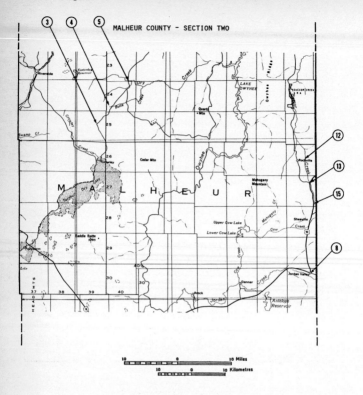

MALHEUR COUNTY – SECTION TWO

10 Miles
10 Kilometres

10. Owyhee area: 4 miles west; turn left on dim trail and go 0.5 mile south to creek.

Petrified wood: good quality. (1-7-13-20)

11. Owyhee area: 8.9 miles west and south on Owyhee Dam Road; 2 miles west of Mitchell Butte; at Niggerhead Rock.

Opal, wood: fine. Petrified wood. (1-3)

12. Rockville area: 1.5 miles east to Succor Creek; collect downstream (north) along both sides of Succor Creek.

Opal, wood: fine. Petrified wood. (1-3-7-13-20)

13. Rockville area: 5 miles south to US-95; turn left and go a short distance to the Succor Creek bridge; upstream (south) along both sides of creek.

Opal, wood: fine. (1-3-7-13-20)

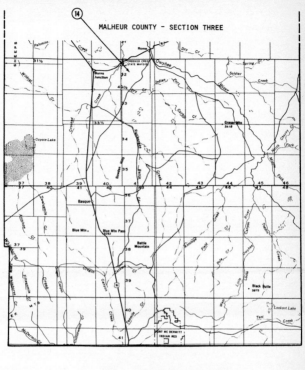

MALHEUR COUNTY — SECTION THREE

14. Rome area: 4.2 miles southwest on US-95 to 1 mile east of rest area and just east of Crooked Creek; turn south and go 1 mile to fork; bear right and go 1 mile to left turn; follow this to end of road.

   Opal, wood. (1-7-13-20)

15. Sheaville area: 3.1 miles north on US-95; collect in Succor Creek Canyon east of highway for 3.5 miles downstream.

   Agate (Thundereggs): to 6″ diameter. Petrified wood, red. (1-7-13-20)

16. Willowcreek area: collect upstream and down along Willow Creek.

   Opal, wood: some very good. (1-3-7-13-20)

MARION COUNTY

1. Detroit area: in Santiam River placers.
   Gold. (17-21)

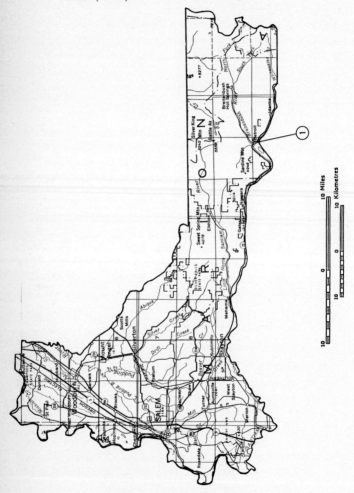

MORROW COUNTY

1. Heppner area: on Opal Butte (aka Peters Butte).
   Hyalite: transparent; translucent. Opal, precious: fine
   gem quality. (1-6-13)

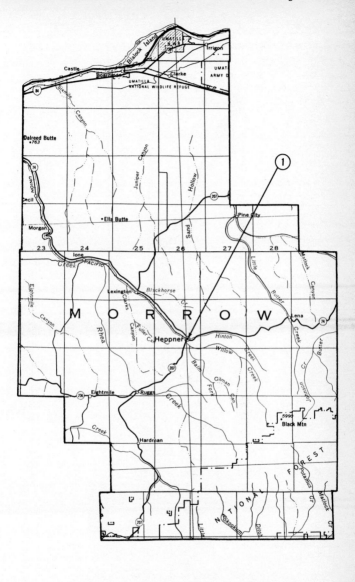

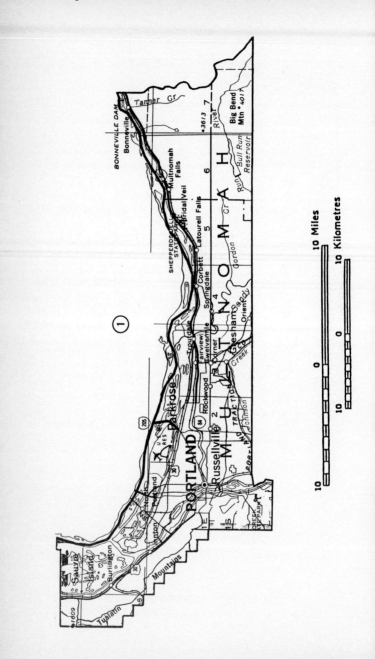

MULTNOMAH COUNTY

1. County-wide in gravel bars of the Willamette River.

   Agate: superb quality. Bloodstone. Jasper. Petrified wood. (3-7-18-20)

POLK COUNTY

1. West Salem area: at the River Bend Sand & Gravel Company works.

   Jasper. Petrified wood. (10)

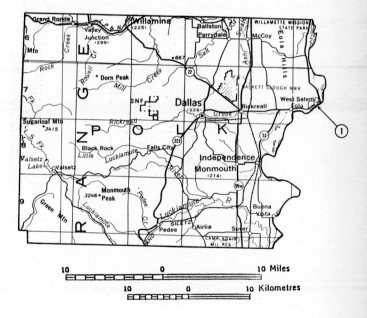

SHERMAN COUNTY

1. Biggs area: 5 miles southeast on US-97; at Fields farm (CF).

   Agate. (1)

2. Rufus area: 1 mile south.
   Agate. (1-13)

3. Wasco area: road cut 1 mile out of town.
   Jasper: scenic. (4)

4. Wasco area: 1.75 miles northwest to broad area centered at junction of road with US-97.
   Wascoite. (1-3-13)

TILLAMOOK COUNTY

1. County-wide along the coastline; on beaches and especially at and near the mouths of all streams.
   Agate. (1)

2. Edwards Butte area.
   Heulandite: fine crystals; pink, white. (5-9-basalt)

3. Lees Camp area: 5.2 miles west; at Cedar Butte.
   Augite: greenish black; crystals to 0.75″. (5-9-basalt)

4. Netarts area: at terminus of Bay-Ocean Road (from Tillamook); at Tillamook Bar.
   Agate: orange. Jasper. (3-18-20)

5. Oceanside area: northward along shore from Maxwell Point.
   Agate: sagenitic. Jasper. Petrified wood. (20)

6. Rockaway area beaches.
   Agate. Jasper. (20)

UNION COUNTY

1. Eagle Cap area: on north slope.
   Agate. (1-13)

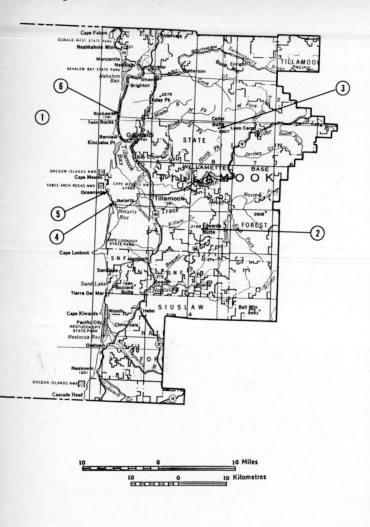

2. Starkey area: at and near the Orofino Mine.
   Agate. (1-2-3)

UNION COUNTY
OREGON

DIVIDED INTO TWO
SEPARATE MAPS
-- EACH MAP
IDENTIFIED BY ITS
NUMBER, AS SHOWN
IN DIAGRAM TO LEFT

UNION COUNTY
SECTION ONE

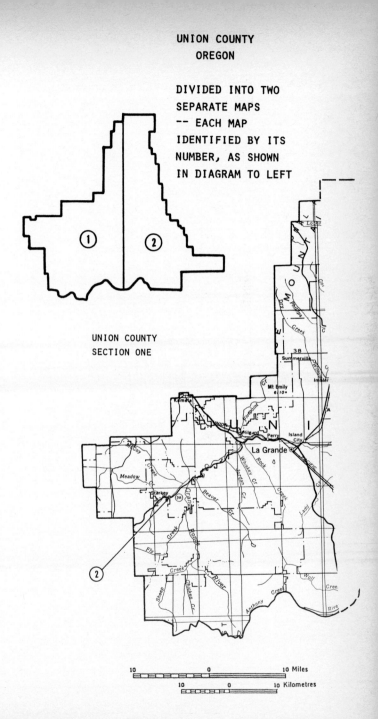

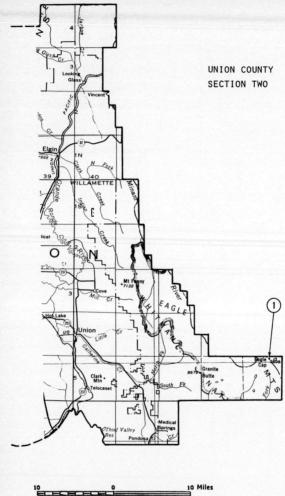

UNION COUNTY
SECTION TWO

10 Miles

10 Kilometres

WALLOWA COUNTY

1. Enterprise area: on Alder Slope of Wallowa Mountain.
   Calcite (Marble var.): black. (1-13-29)

2. Imnaha area: 9.5 miles downstream (northeast) on the Imnaha River to the mouth of Horse Creek; collect upstream
   for 2 miles on the east bank of Horse Creek.
   Agate. (1-3-13-20)

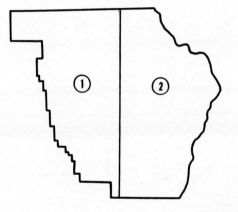

WALLOWA COUNTY
OREGON

DIVIDED INTO TWO
SEPARATE MAPS ON
THE FOLLOWING TWO
PAGES -- EACH MAP
IDENTIFIED BY ITS
NUMBER, AS SHOWN
IN DIAGRAM

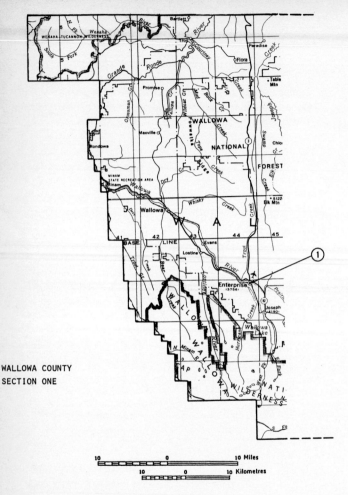

WALLOWA COUNTY
SECTION ONE

10 ———— 0 ———— 10 Miles

10 ———— 0 ———— 10 Kilometres

WASCO COUNTY

1. Antelope area: all around town for several miles; mostly
   private area: some with CF.

   Agate (Thundereggs): to 6″ diameter. Agate, iris:
   slabs to 3″ × 6″; rainbow colors. Agate, moss: nodules
   to enormous sizes (into the tons). (1-3-7-20-39)

WALLOWA COUNTY
SECTION TWO

10 ⌐ 0 ⌐ 10 Miles
10 ⌐ 0 ⌐ 10 Kilometres

2. Antelope area: 0.3 mile east on SR-218 to bridge; after crossing bridge watch for Antelope Bed sign and turn south there (CF).

    Agate (Thundereggs): to 5″ diameter. Agate, moss. Amethyst. (1)

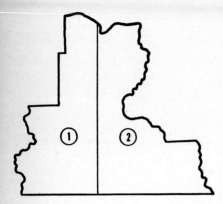

WASCO COUNTY
OREGON

DIVIDED INTO TWO
SEPARATE MAPS ON
THE FOLLOWING TWO
PAGES -- EACH MAP
IDENTIFIED BY ITS
NUMBER, AS SHOWN
IN DIAGRAM TO LEFT

3. Antelope area: 1 mile east on SR-218; then 0.25 mile south; at the Wasco County Road Quarry.

   Jasper: red. (12)

4. Antelope area: 1.25 miles east on SR-218; at Brown Ranch (CF).

   Jasper: red. (1-3-13)

5. Antelope area: 3.1 miles east on SR-218 to side road at right leading toward Ashwood (Jefferson County); take it 2.7 miles south; collect on both sides of the road just north of the Jefferson County border.

   Agate (Thundereggs): to 5″ diameter. Agate, moss: green. (1-3-13)

6. Clarno area: at the Oregon State Quarry and surrounding area.

   Agate. Jasper. Petrified wood. (1-3-12-13)

7. Mosier area: in the vicinity of Wasco Lake.

   Petrified wood: includes silicified pine cones. Wascoite: with marine fossil inclusions. (3-7-20)

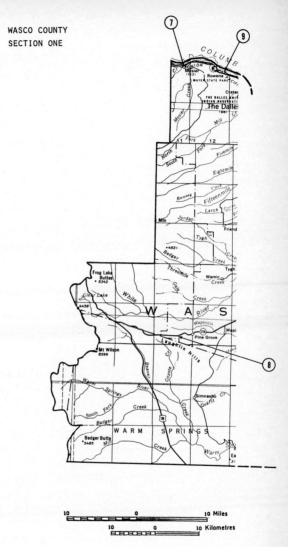

WASCO COUNTY
SECTION ONE

10   0   10 Miles

10   0   10 Kilometres

8. Pine Grove area: 2.1 miles west on SR-216; then south in the Laughlin Hills to Sunflower Flats.
   Agate. (1-3-13-39)

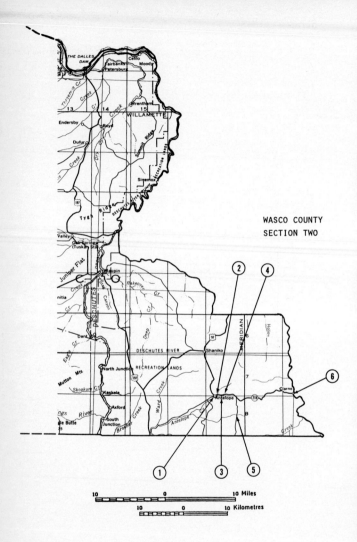

WASCO COUNTY
SECTION TWO

10 Miles | 0 | 10 Miles
10 | 0 | 10 Kilometres

9. Rowena area: southeast toward Chenoweth; at Johnson Ranch.

Petrified wood. (1-13)

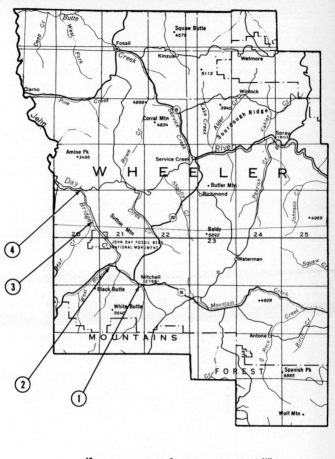

WHEELER COUNTY

1. Mitchell area.

   Stilbite: fine; white; crystals to 1". (1-13)

2. Mitchell area: 4.5 miles west on US-26 to West Branch of
   Bridge Creek; collect downstream on both sides to the
   south boundary of John Day Fossil Beds National Monument.

   > Chert, banded (aka Morrisonite): fine; very colorful.
   > Fossils, flora and fauna: some excellent specimens,
   > including silicified bone. (1-2-3-7-20)

3. Mitchell area: 4.6 miles west on US-26 to side road north
   leading to John Day Fossil Beds National Monument; take
   this road 6.25 miles north to 1.2 miles north of the north
   boundary of the national park; then 0.3 mile west to where
   Bear Creek enters West Branch of Bridge Creek; collect
   downstream (north) to the creek's mouth at the John Day
   River.

   > Chert, banded (aka Morrisonite): fine; very colorful.
   > (1-2-3-7-13-20)

4. Mitchell area: 4.6 miles west on US-26 to side road north
   leading to John Day Fossil Beds National Monument; take
   this road 11.2 miles, through the national monument and
   to the John Day River.

   > Chert, banded (aka Morrisonite): fine; very colorful.
   > (1-2-3-7-13-20)

SOUTH DAKOTA

Statewide total of 90 locations in the following counties:

Bon Homme (1)	Fall River (12)	Pennington (22)
Butte (2)	Harding (1)	Shannon (1)
Campbell (1)	Jackson (5)	Stanley (1)
Charles Mix (1)	Jones (1)	Todd (1)
Corson (2)	Lawrence (7)	Yankton (1)
Custer (23)	Meade (4)	Ziebach (1)
Douglas (1)	Minnehaha (2)	

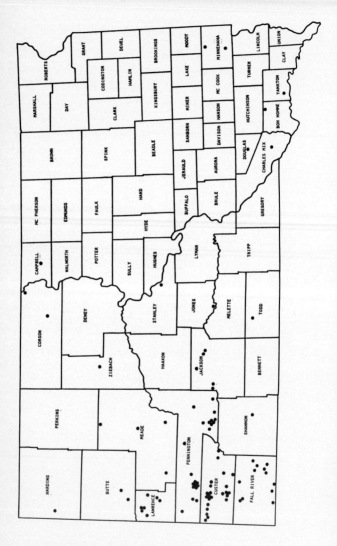

COLLECTING SITES IN SOUTH DAKOTA

(FOR SPECIFIC DETAILS ON THESE
LOCATIONS, SEE THE FOLLOWING
COUNTY MAPS AND RELATED TEXT)

BON HOMME COUNTY

1. Scotland area: collect in broad area west and south-south-west, north of a line running from Scotland to where SR-46 enters Charles Mix County.

    Selenite: roses. (1-13)

BUTTE COUNTY

1. Fruitdale area: 1 mile west on US-212 to right turn at a sign for the Belle Fourche Reservoir, Public Access Road; 3.5 miles north on gravel road to Stony Point.

    Barite, golden: crystals. Fossils, marine: especially ammonites and brachiopods. Petrified wood. Selenite: fishtail crystals. (1-5-8-13-20)

2. Nisland area: 3.3 miles west on US-212 to side road right leading to Belle Fourche Dam; follow to dam, cross the dam, and then turn left on dirt road; take this to the collecting area along the shoreline.

Barite, golden: crystals. Fossils, marine: especially ammonites and brachiopods. Petrified wood. Selenite: fishtail crystals. (1-5-8-13-20)

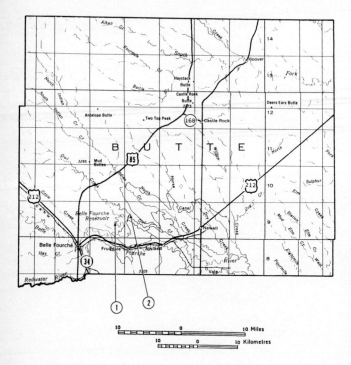

CAMPBELL COUNTY

1. County-wide on hillsides, knobs, and knolls and in arroyos, streambeds, washes, streambanks, and related areas in the Missouri River valley.

Opal, wood: fine. (1-3-13)

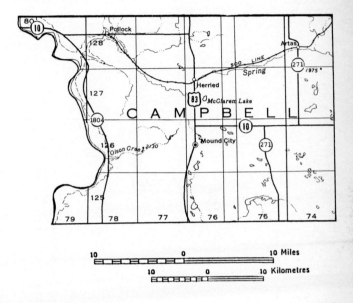

10   0   10 Miles

10   0   10 Kilometres

CHARLES MIX COUNTY

1. Wagner area: in road cuts and stream cuts.

    Selenite: very fine; bladed crystals in rose clusters; honey-colored. (4)

CORSON COUNTY

1. Kenel area: northward to the North Dakota border and southward to the Wakpala area; on hillsides and knobs, at mouths of streams, and along banks of the Oahe Reservoir of the Missouri River valley.

    Opal, wood: fine. (1-3-17-20)

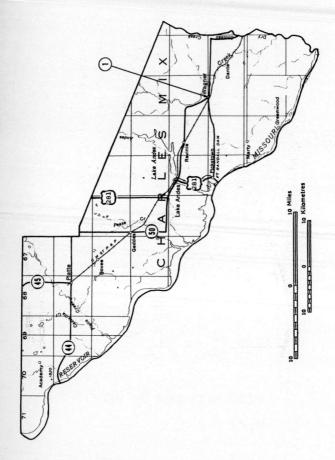

2. Little Eagle area: upstream and down along the shores of the Grand River.

    Opal, wood: fine. Petrified wood. (1-3-7-18-20)

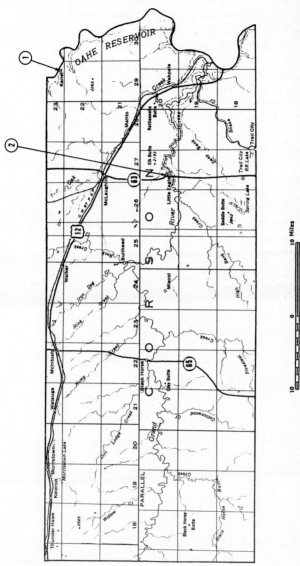

CUSTER COUNTY

1. Buffalo Gap area: along side roads and trails within a 10-mile radius of town.

    Agate, fortification (Fairburn var.). (1-13)

2. Custer area: broad region of beds surrounding town; heavily collected area—be sure to check sites of any new excavations of any kind.

    Agate, fortification (Fairburn var.) (1-3-4-13)

3. Custer area: east on US-16 to Custer State Park; take turn-off toward State Game Farm; go just beyond cattle guard and collect in hills to the right.

    Agate, fortification (Fairburn var.). (1-13)

4. Custer area: in Hell Canyon; accessible by road from US-16.

    Agate: geodes to 12″ diameter. Agate, fortification (Fairburn var.): nodules to 10″ diameter. (5-sandstone -14-29-39)

5. Custer area: in outcroppings of surrounding hills.

    Quartz, rose. (4-5-14-19-granite -29)

6. Custer area: just south of State Game Farm on US-16.

    Agate, fortification (Fairburn var.). Chalcedony. Jasper. (1-3)

7. Custer area: northeast; along SR-87 and on secondary road behind Mount Rushmore, running from US-16 to US-16A.

    Quartz, rose. (1-4-13-29)

8. Custer area: 4 miles west; in Elephant Gulch.

    Garnet, almandine: red; flawless crystals; also in nodules. Gold. (17)

9. Custer area: 5 miles southeast; at the Bull Moose Mine.

    Barbosalite: cabinet quality. Pyrite: bright brassy crystals. (2)

10. Custer area: 6 miles southeast; in canyon on branch of French Creek; at the Scott Rose Quartz Quarry.

> Lithiophilite. Quartz, rose: very fine; in milky quartz core of 50′ × 1200′ pegmatite. (12-16)

11. Custer area: 7 miles west on US-16 to side road north at sign saying "Deer Camp 9 Miles"; go 0.25 mile north; collect on slopes and at base of Tin Mountain and in stream below.

> Garnet, almandine. Tourmaline. (1-7-13-27-mica -29)

12. Custer area: 14 miles west on US-16 past Jewel Cave National Monument to Tepee Canyon; turn north to diggings.

> Agate: clear. Agate, fortification (Fairburn var.). Beryl. Garnet, almandine. Lepidolite. Quartz, rose. Staurolite. Tourmaline. (1-8-13-14-29)

13. Custer area: 15 miles west on US-16 to 1 mile past campground on left; take side road right going up canyon; collect on crest of hill to left and also up canyon to small wash and collect there on hill to left.

> Agate, fortification (Fairburn var.). (5-limestone)

14. Custer area: 17 miles west on US-16; take second right after passing campground on left; in the area of the junction of this logging road.

> Agate, fortification (Fairburn var.). (1-13)

15. Fairburn area: 3.5 miles east on SR-79 to fork; bear left and follow French Creek 5.5 miles to McDermand Ranch; continue to picnic area: enter it, drive across creek, and go about 1 mile to collect on gravelly hills.

> Agate, fortification (Fairburn var.). (1-3-13-39)

16. Fairburn area: 6.4 miles southwest to dry bed of Lame Johnny Creek; collect in both directions in streambed.

> Agate, fortification (Fairburn var.). Jasper. (1-7-13-40)

17. Fairburn area: 15 miles east on French Creek Road to Buffalo Gap National Grasslands Picnic Ground (formerly the 4-H Wranglers Picnic Ground); turn north through picnic ground; go through dry creekbed and follow road-trail toward buttes for 1.1 miles around, up and over a hill to the vast Fairburn Agate Beds; entire route toward buttes is agate beds; good surface material is scarce due to heavy collecting, but excellent finds can still be made by probing and digging.

> Agate, fortification (Fairburn var.). Jasper: quality; red, yellow. Petrified wood: including palmwood. (1-13-39)

18. Pringle area: heading west; along all intersecting roads and trails north and south, as far south as Minnekahta (Fall River County).

> Agate, fortification (Fairburn var.). (1-3-13)

19. Pringle area: in broad area to the north and northwest.

> Agate, fortification (Fairburn var.). (1-13-39)

20. Pringle area: northeast; adjacent to south border of Custer State Park.

> Agate, fortification (Fairburn var.). (1-13-39)

21. Pringle area: 2.8 miles east on US-18/385; less than 200' north of road.

> Quartz, rose. (24)

22. Pringle area: 3 miles southwest of Wind Cave National Park; in the area adjacent to Onyx Cave.

> Agate, fortification (Fairburn var.). (1-13-39)

23. Sanator area: 2 miles south on US-385 to fork; bear right to ridge; at the White Elephant Mine.

> Quartz, rose. (2)

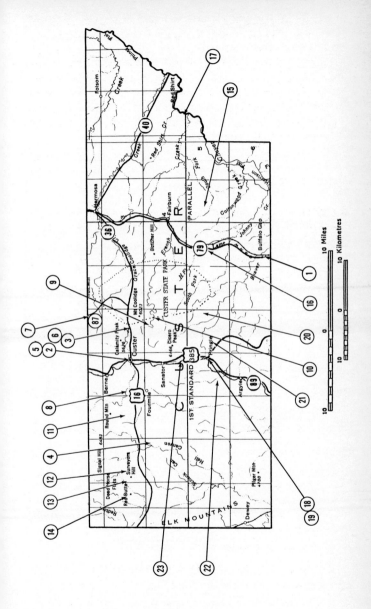

DOUGLAS COUNTY

1. County-wide in road cuts, stream cuts, and especially new
   excavations, particularly in the areas of Corsica and Ar-
   mour.

   Selenite: fine; bladed crystals in rose clusters. (4)

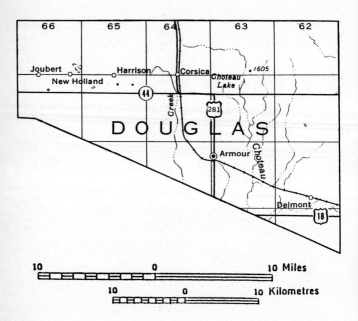

FALL RIVER COUNTY

1. Ardmore area: southward to the Nebraska border; then
   east and west along the state line.

   Agate, fortification (Fairburn var.). Jasper. (1-13)

2. Ardmore area: 7 miles east on side road to the Flemming Ranch; then east-southeast about 4 miles to an area of breaks on the Nebraska border.

>   Agate, fortification (Fairburn var.). Chalcedony: dendritic. Jasper: scarlet, lemon, green, lilac. Petrified wood. (1-13-39)

3. Ardmore area: 7 miles east on dirt road; at Sugarloaf Butte.

>   Agate. Jasper. (1-13)

4. Hot Springs area: in Fall River Canyon.

>   Petrified wood. (1-13)

5. Minnekahta area: 1.8 miles south; go through farmyard on Forest Service trail leading to Parker Peak Ranger Station; at the top, take left fork; follow main trail, bearing right; go past dam and through three gates to broad flat; collect on the flat and in hills to the right.

>   Fossils: cycads; silicified; some whole and very well defined. Petrified wood: including palmwood. (1-13)

6. Minnekahta area: 2 miles south; 1.5 miles east of US-18; immediate area of Parker Peak.

>   Fossils: cycads: silicified. (13-14)

7. Oelrichs area: south on US-385 to first side road east; take it 1 mile to side road south; go south 2 miles and then continue following it at sharp turn to the east and through series of sharp turns as road gets progressively worse; when main trail turns directly south, turn north on dim trail and go 1.5 miles to breaks.

>   Agate, fortification (Fairburn var.). Fossils: cycads and other flora; silicified. Petrified wood. (1-13-39)

8. Oelrichs area: south from center of town to side road left; go east 1 mile to side road right; follow latter 2 miles to farm; then go east on a faint trail; then north 1 mile; finally go west to buttes.

>   Agate, fortification (Fairburn var.). (1-13)

9. Oelrichs area: 1 mile east to side road south; go 0.25 mile south to buttes.

> Agate, fortification (Fairburn var.): yellow, black and white. Calcite. Concretions: some containing fluorescent materials. Fossils: cycads and other flora; some silicified. Petrified wood. (1-13-39-41)

10. Oelrichs area: 1.5 miles east from junction of US-18 and US-385 to gravel side road on right; 3 miles south (2 miles on gravel, then dirt) to dirt side road on left; 0.7 mile east on latter to wire gate; continue through and go another 3.6 miles, at which point you are in the midst of an expansive agate bed.

> Agate, fortification (Fairburn var.). (1-3-6-13)

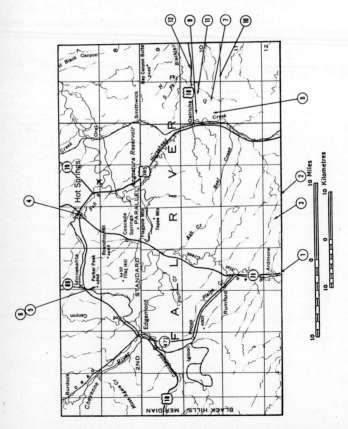

11. Oelrichs area: 4 miles east to ranch entry at right; go south 0.5 mile; on north and east slopes of butte.

> Agate, fortification (Fairburn var.). (1-13)

12. Oelrichs area: 7.4 miles east; in area of breaks.

> Agate, fortification (Fairburn var.). Petrified wood. (1-13-39)

HARDING COUNTY

1. Camp Crook area: upstream on both sides of the Little Missouri River, as well as all its tributaries, to the Montana border.

> Chalcedony: nodules to 7″ diameter; translucent; dendritic. (1-3-7-13-18-20)

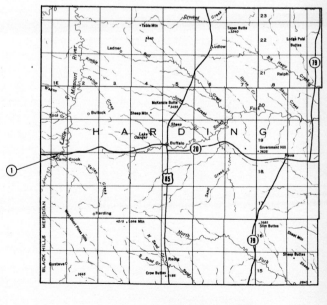

JACKSON COUNTY

1. Interior area: in eroded areas of widespread area around town.

    Agate, fortification (Fairburn var.). Puddinstone. (1-13-39)

2. Interior area: west on SR-44 to 0.75 mile west of where pavement ends; along both sides of the continuing gravel road.

    Agate, fortification (Fairburn var.). (1-13)

3. Kadoka area: from junction of I-90 Byp and SR-73, go 1.5 miles south to dirt side road on right; 4.8 miles west on latter to fork (NOTE: impassable in wet weather; passable in dry weather for two-wheel drive, but four-wheel drive is recommended); bear right at fork and go 8.5 miles to another fork; bear right again and go 0.7 mile; collect on both sides of road.

    Agate (Bubblegum and Prairie vars.). Agate, fortification (Fairburn var.). Petrified wood. (1-13-39)

4. Kadoka area: in eroded areas widely surrounding town.

    Agate, fortification (Fairburn var.). Puddinstone. (1-13-39)

5. Philip Junction area: 12 miles east to dirt trail on right; 2 miles south on latter; collect toward southwest in Buffalo Gap National Grassland.

    Agate, fortification (Fairburn var.). Beryl. Calcite. Fossils, vertebrate: mammal bones and teeth. Jasper. Petrified wood. Quartz, rock. (1-13-39)

JONES COUNTY

1. Westover area: 2.3 miles south on US-83 to bridge over the White River.

    Chalcedony. (36)

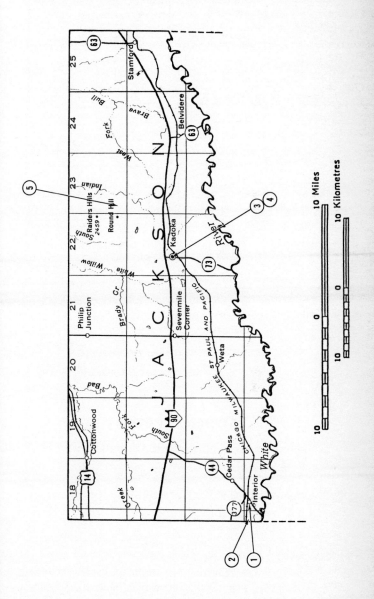

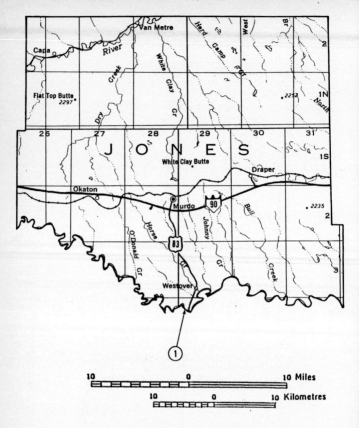

LAWRENCE COUNTY

1. Central City area: at commercial operations (CF).
   Gold. (2-17-21)

2. Deadwood area: in the Black Hills.
   Chert: cutting quality; red, gray. Ferberite. Heuber-
   nite. (5-limestone -16-19)

3. Spearfish area: in the Black Hills.
   Gypsum (Alabaster var.): white, gray; often stained or
   mottled from iron oxide; found in so-called red beds.
   (19-39)

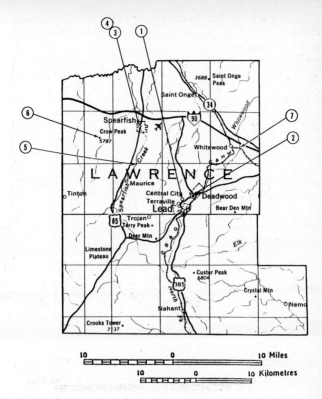

4. Spearfish area: in Spearfish Canyon.

   Amethyst: geodes to 6″ diameter. Chalcedony: geodes
   to 6″ diameter. (4-14)

5. Spearfish area: south on US-85 to Savoy; then downstream
   in Spearfish Canyon, along both sides of Spearfish Creek
   and upstream on both sides of both forks of the same
   stream.

   Amethyst: geodes to 9″ diameter. (1-7-13-20)

6. Spearfish area: 4.6 miles west-southwest; on Crow Peak.

   Calcite (Onyx var.): black. (1-13)

7. Whitewood area: 1.15 miles east to Whitewood Creek; upstream and down on both sides of the creek.

> Amethyst: geodes to 11″ diameter; nodules to 8″ diam-
> eter. Chalcedony: geodes to 12″ diameter. (1-4-7-13-14-
> 20)

MEADE COUNTY

1. County-wide.

> Barite: rich, dark, amber-colored; prismatic crystals;
> in barite concretions. (1-3-14-41)

2. Elm Springs; 4.2 miles south to Elk Creek River; collect
   upstream and down on both sides; also in banks on both
   sides.

> Barite: fine; clear golden crystals in clusters; also
> geodes to 10″ diameter; irregular shapes; interiors
> lined with yellow calcite drusy from which project fine
> barite crystals to 2″ long; crystals are pale yellow to
> dark brown. Calcite: yellow crystals in clusters; also
> as fine drusy within barite geodes. (1-5-Pierre shales
> -7-8-13-14-20-22-28-38-41)

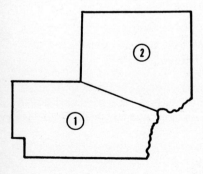

MEADE COUNTY
SOUTH DAKOTA

DIVIDED INTO TWO
SEPARATE MAPS ON
THE FOLLOWING TWO
PAGES -- EACH MAP
IDENTIFIED BY ITS
NUMBER, AS SHOWN
IN DIAGRAM TO LEFT

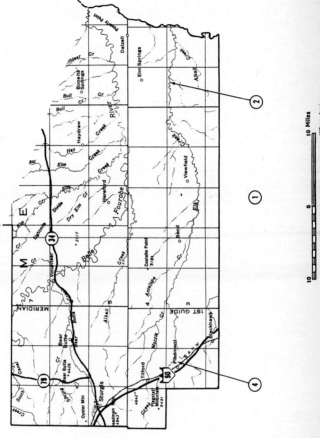

MEADE COUNTY – SECTION ONE

3. Maurine area: 4.5 miles east-southeast; atop Fox Ridge.
   Agate, moss. (5-limestone -19)

4. Piedmont area: 0.75 mile east.
   Fossils: belemnites; silicified. Selenite: fine, clear crystals. (1-13)

MEADE COUNTY - SECTION TWO

MINNEHAHA COUNTY

1. Dell Rapids area: upstream and down in the Big Sioux River.

    Agate. Chalcedony (aka Sioux Falls jasper): in conglomerates. (7-10-40)

2. Sioux Falls area: upstream and down in the Big Sioux River.

 Agate. Chalcedony (aka Sioux Falls jasper): in conglomerates. Jasper. (7-10-40)

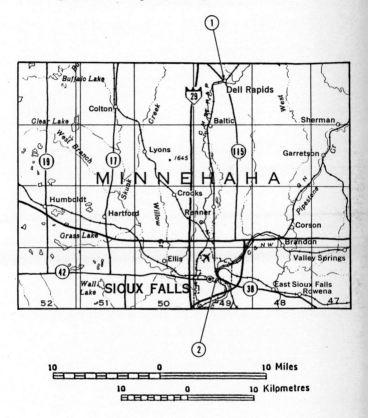

PENNINGTON COUNTY

1. Bear Creek Butte area: in the Badlands.
 Amethyst. (1-26)

2. Hill City area mine dumps.
 Agate, fortification (Fairburn var.). Apatite. Beryl. Kunzite. Lepidolite. Zircon. (2)

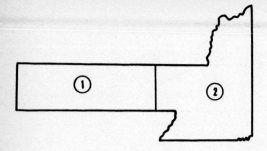

PENNINGTON COUNTY
SOUTH DAKOTA

DIVIDED INTO TWO
SEPARATE MAPS ON
THE FOLLOWING TWO
PAGES -- EACH MAP
IDENTIFIED BY ITS
NUMBER, AS SHOWN
IN DIAGRAM

3. Hill City area: surrounding hills in a wide radius.
   Quartz, rose. (5-14-19-granite -29)

4. Imlay area: east on SR-44 to first road south; at White
   River.
   Agate: black, white. (20)

5. Keystone area.
   Autunite. Beryl: golden. Triphylite. (16-several)

6. Keystone area: along gravel road that parallels the US-
   16A tunnel between town and junction with US-16.
   Garnet. Staurolite. Tourmaline. (1-13-14-24-29)

7. Keystone area: at Robert Ingersoll Mine.
   Lepidolite. Tourmaline: opaque; colored. Triphylite.
   (2)

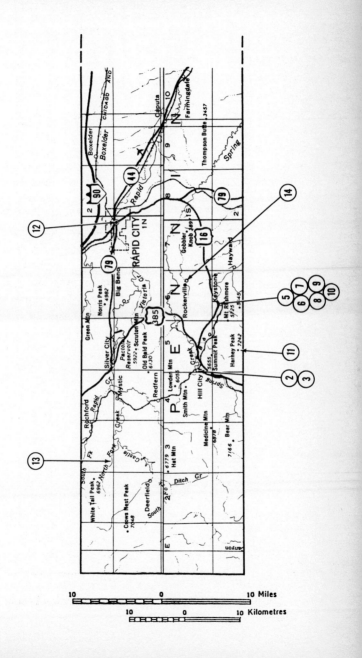

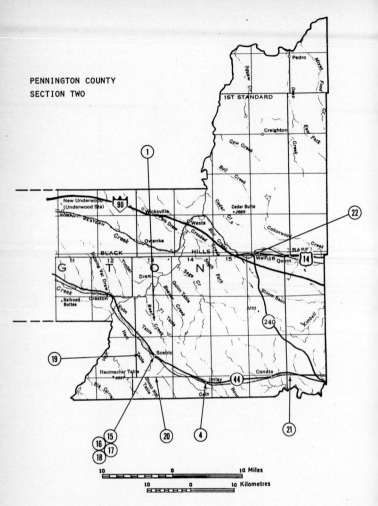

PENNINGTON COUNTY
SECTION TWO

8. Keystone area commercial operations on Battle River
   tributaries.
      Gold. (2-17-21)

9. Keystone area mine dumps.
      Tourmaline: blue, red. (2)

10. Keystone area: near Mount Rushmore on both sides of a gravel road connecting US-16A and SR-87.

Quartz, rose. Triphylite. (16)

11. Keystone area: west to Harney Peak; on slopes and at base.

Tourmaline: green; poor to fair crystals; to 12″ thick. (1-13-14-16)

12. Rapid City area: east; in Boxelder Creek and adjacent grounds.

Amethyst. Quartz, rock: rutilated. (1-7-13-20-26)

13. Rochford area: 4 miles southwest to North Fork; collect upstream and down.

Gold. (17)

14. Rockerville area commercial operations on US-16A (CF).

Gold. (2-17-21)

15. Scenic area: in eroded areas widely surrounding town.

Agate, fortification (Fairburn var.). Jasper. Petrified wood. Puddinstone. (1-13-39)

16. Scenic area: northeast to Hay Draw formation; especially in creek banks.

Agate, fortification: to 5″ diameter. (1-5-clay ironstone -7-14-20)

17. Scenic area: southeast; collect along south side of road all the way to Imlay; especially abundant just southwest of Imlay.

Chalcedony: geodes to 11″ diameter; lined with crystals of dolomite. Dolomite: crystals lining interiors of chalcedony geodes. Selenite: bladed crystals; transparent; tan. (1-7-13)

18. Scenic area: west to Cheyenne River breaks; in area of the Hedlun cabin.

Agate, fortification (Fairburn var.). (1-13)

19. Scenic area: 2 miles west on SR-40 to closest point to Hart Table Buttes (red hills) on right; walk toward them, collecting en route, and at their base.

> Agate: nodules to 3″ long; black with white cores. Agate, fortification (Fairburn var.). Jasper. (1-8-13-14)

20. Scenic area: 3 miles south; in a 2-mile belt bisected by the road for 1 mile.

> Agate, fortification (Fairburn var.). Petrified wood. Quartz, rose. (1)

21. Scenic area: 24.5 miles east on SR-44; 7.75 miles east of Conata to MM-113; turn left onto trail and go north to where there is a sign between two rises; this is the beginning of the extensive Conata Agate Beds.

> Agate (Prairie var.). Agate, fortification (Fairburn var.). (1-13-39)

22. Wall area: collect along roads paralleling I-90, all the way to Rapid City.

> Barite: fine; transparent; rich brown crystals lining barite concretions. Calcite: fine; yellow; crystals and drusy within concretions. (1-4-13-39-alkali -41)

SHANNON COUNTY

1. Porcupine area: along both sides of Porcupine Creek; collect upstream (southeast) to headwaters and downstream (north-northwest) to Sharps Corner.

> Agate, fortification (Fairburn var.). Chalcedony: blue. (1-7-13-20-29)

STANLEY COUNTY

1. Fort Pierre area: upstream on both sides of the Bad River, all the way to Philip (in Haakon County).

> Fossils, invertebrate: ammonites, baculites, nautiloids, scaphites; all well silicified. (1-7-13-14-41)

TODD COUNTY

1. Parmelee area: 6.85 miles east to the South Fork of the
   White River (aka Little White River); upstream and down
   on both sides and on adjacent hillsides.

   Fossils, invertebrates: wide variety; well silicified;
   very abundant. Petrified wood (Black opal var.).
   (1-7-13-20)

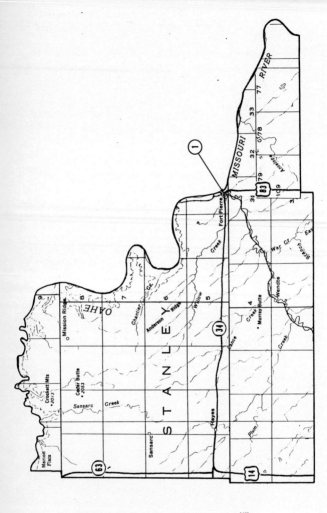

10 0 10 Miles

10 0 10 Kilometres

YANKTON COUNTY

1. Yankton area: 4 miles west-southwest on Gavin Point Road; at abandoned brickyard on right; in the yard area and especially in clay pits.

    Selenite: loose crystals and bladed rose clusters; transparent; honey-colored. (1-12-13-35)

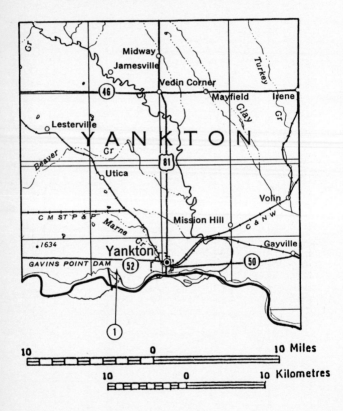

ZIEBACH COUNTY

1. Dupree area: southwest to Rattlesnake Butte; collect on crest and slopes of south butte.

    Calcite, sand: excellent crystals to 8″ long and 2″ thick; well terminated. (1-13-14)

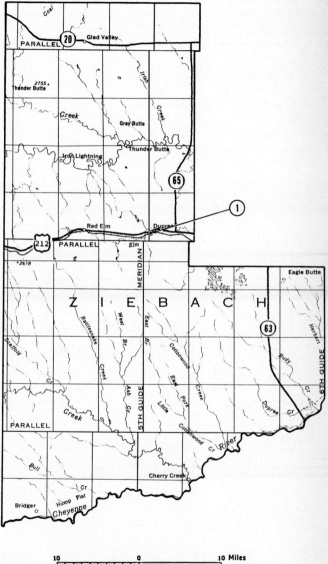

10 Miles

10 Kilometres

WASHINGTON

Statewide total of 146 locations in the following counties:

Asotin (2)	Island (1)	San Juan (1)
Benton (3)	Jefferson (4)	Skagit (5)
Chelan (3)	King (5)	Skamania (1)
Clallam (6)	Kittitas (21)	Snohomish (1)
Clark (4)	Klickitat (8)	Spokane (3)
Cowlitz (8)	Lewis (11)	Stevens (2)
Douglas (3)	Lincoln (3)	Thurston (4)
Ferry (8)	Okanogan (6)	Wahkiakum (1)
Franklin (2)	Pacific (3)	Whitman (3)
Grant (9)	Pend Oreille (3)	Yakima (6)
Grays Harbor (5)	Pierce (1)	

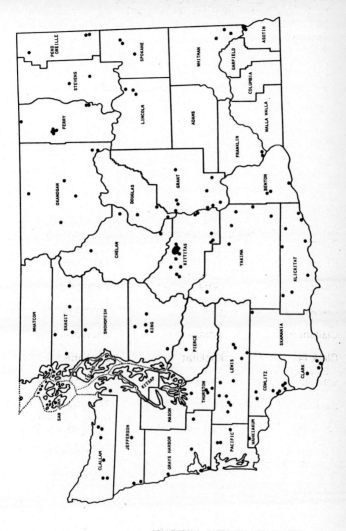

COLLECTING SITES IN WASHINGTON

(FOR SPECIFIC DETAILS ON THESE
LOCATIONS, SEE THE FOLLOWING
COUNTY MAPS AND RELATED TEXT)

ASOTIN COUNTY

1. Clarkston area.
   Gold. (3-17)

2. Clarkston area: in Snake River sand bars.
   Gold. (17-21)

BENTON COUNTY

1. Paterson area: in the Horse Heaven Hills Range where it meets the Columbia River.

   Opal, wood: fine; cell structure and annular rings visible. (1)

2. Prosser area: in open range land.

   Opal, wood: fine; cell structure and annular rings visible. (1-13)

3. Prosser area: in canyons of the Horse Heaven Hills.

   Opal, wood. (31)

CHELAN COUNTY

1. Chelan Lake area: generally in lake area.
   Molybdenite: excellent crystals. (5-quartz -29)

2. Lucerne area: upstream along both sides of Railroad Creek to Holden.
   Agate. Jasper. (1-3-7-13-18-20)

3. Wenatchee area: just outside town in Number One Canyon.
   Calcite (Onyx var.): white; translucent; abundant. (1-3)

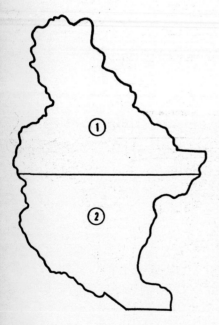

CHELAN COUNTY
WASHINGTON

DIVIDED INTO TWO
SEPARATE MAPS
-- EACH MAP
IDENTIFIED BY ITS
NUMBER, AS SHOWN
IN DIAGRAM TO LEFT

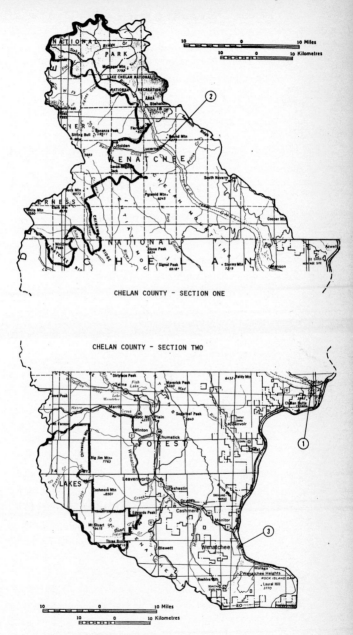

CHELAN COUNTY – SECTION ONE

CHELAN COUNTY – SECTION TWO

CLALLAM COUNTY

1. County-wide; along virtually every gravel shoreline of the Olympic Peninsula, seaside and inland.

   Chalcedony: high quality. (3-20)

2. Fairholm area: at the mouth of the Quillanute River.

   Jasper: orbicular. (7-20)

3. Fairholm area: 1 mile northwest of the west end of Lake Crescent.

   Jasper: orbicular; boulders. (1-13)

4. Fairholm area: 6.5 miles west on US-101.

   Jasper: orbicular. (1-13)

5. Forks area: 0.75 mile north on SR-101 to the Calawah River; collect upstream and down on both sides.

   Jasper: orbicular. (7-20)

6. Forks area: 2 miles north on SR-101 to the Soleduck River; collect upstream and down on both sides.

   Jasper: orbicular. (7-20)

CLALLAM COUNTY
WASHINGTON

DIVIDED INTO TWO
SEPARATE MAPS ON
THE FOLLOWING TWO
PAGES -- EACH MAP
IDENTIFIED BY ITS
NUMBER, AS SHOWN
IN DIAGRAM TO LEFT

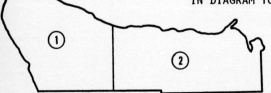

CLALLAM COUNTY
SECTION ONE

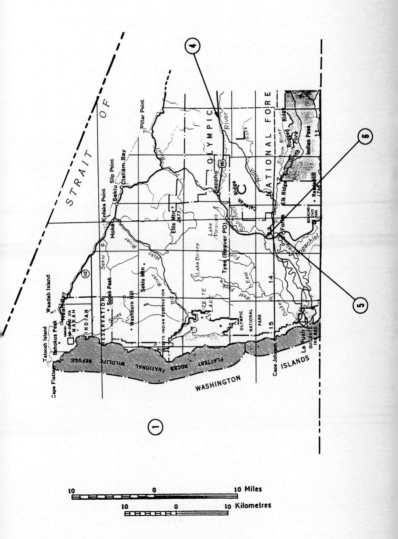

CLALLAM COUNTY - SECTION TWO

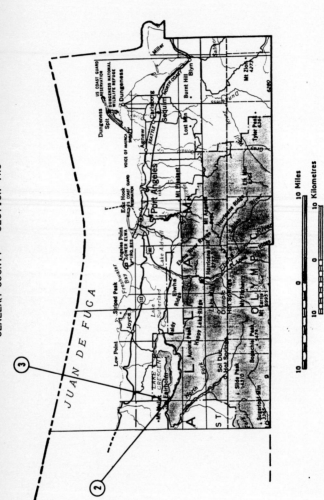

CLARK COUNTY

1. Brush Prairie area streambeds.
      Gold. (17-21)

2. Camas area: in Columbia River sand bars.
      Gold. (17)

3. Washougal area.
      Agate, moss. Amethyst. (1-13)

4. Washougal area: 2.5 miles northeast.
      Agate, moss. Amethyst. (1-13)

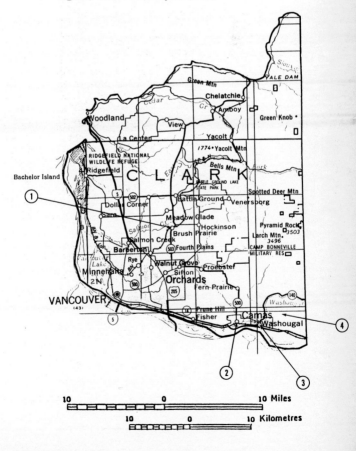

COWLITZ COUNTY

1. Castle Rock area: east on SR-504 to Kid Valley State Park; then east to Beaver Creek.

    Carnelian. (7-20)

2. Kalama area: east side of the Columbia River.

    Agate. (1-7-13-18-20)

3. Kalama area: to Gilmore; then south on Sightly Road, which turns east; turn right at first road south; at Walter Smith farm (CF).

    Bloodstone. Carnelian. (1-13)

4. Kalama area: 1.5 miles northeast on Cemetery Road; in cliff.

    Agate: geodes to 6.5″ diameter. Agate (Thundereggs): to 5″ diameter. (1-8-13-14)

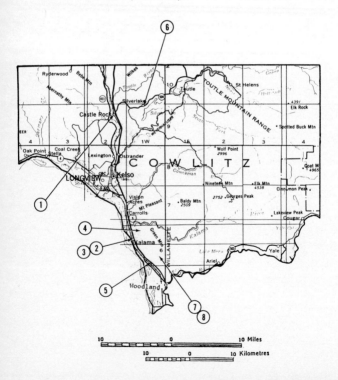

5. Kalama area: 4 miles southeast on Green Mountain Road to Johnson Road; turn northeast and go 0.25 mile.

   Agate: geodes to 6″ diameter. Agate (Thundereggs): to 5″ diameter. (1-13)

6. Silverlake area.

   Agate. Petrified wood. (1-7-13-20)

7. Woodland area: 5 miles north on US-99 (I-5) to Cloverdale; east on Todd Road to Cloverdale Road; then north and east to Green Mountain Road; in Cloverdale Creek.

   Carnelian. (7-20)

8. Woodland area: 5 miles north on US-99 (I-5) to Cloverdale; 4 miles east; in broad area.

   Agate: geodes to 14″ diameter; many amethyst-lined. Carnelian. Chalcedony. (1-3-13-39)

DOUGLAS COUNTY

1. Palisades area: from west of road at Moses Coulee south to Appledale.

   Opal, wood. Petrified wood: large logs and limbs. (1-13)

2. Rock Island area: 4.8 miles southeast on SR-28; on east side of Rock Island Dam.

   Pyrite: fine crystals. (37)

3. Waterville area.

   Opal, common: green. (1)

FERRY COUNTY

1. Covada area: at the Longstreet Mine.
   Stibnite. (2)

2. Covada area mines.
    Stibnite. (2)

3. Danville area mines.
    Gold. (2)

4. Republic area: at the Lone Pine Mine.
    Gold. Silver. (2)

5. Republic area: at the Quilp Mine.
    Gold. Silver. (2)

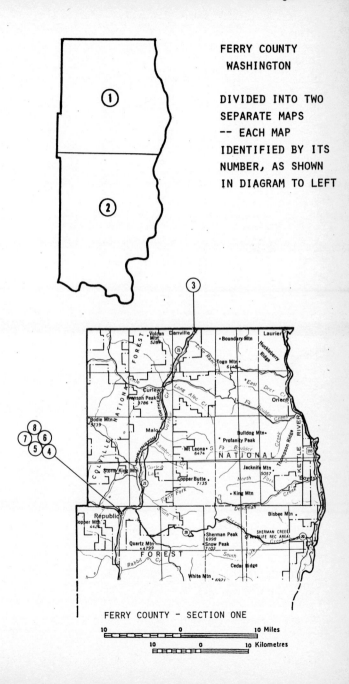

FERRY COUNTY
WASHINGTON

DIVIDED INTO TWO
SEPARATE MAPS
-- EACH MAP
IDENTIFIED BY ITS
NUMBER, AS SHOWN
IN DIAGRAM TO LEFT

FERRY COUNTY - SECTION ONE

10   0   10 Miles

10   0   10 Kilometres

FERRY COUNTY - SECTION TWO

6. Republic area: at the Surprise Mine.
   Gold. Silver. (2)

7. Republic area: at the Tom Thumb Mine.
   Gold. Silver. (2)

8. Republic area: upstream and down along Granite Creek.
   Gold. (3-17-18)

FRANKLIN COUNTY

1. Pasco area: downstream along the east side of the Columbia River; from bluffs to water.

    Agate. (1-5-bluffs -8-13-20)

2. Pasco area: to Ringgold area; on elevated gravel bars of the Columbia River.

    Agate: abundant; as pebbles. (3-18)

GRANT COUNTY

1. Beverly area: 6 miles upstream along the Columbia River to the wooden bridge over Crab Creek; north bank of creek and hills.

    Petrified wood. (1-7-13-20)

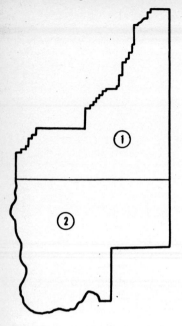

GRANT COUNTY
WASHINGTON

DIVIDED INTO TWO
SEPARATE MAPS ON
THE FOLLOWING TWO
PAGES -- EACH MAP
IDENTIFIED BY ITS
NUMBER, AS SHOWN
IN DIAGRAM TO LEFT

2. Corfu area.

Petrified wood: entire forest; limbs and logs. (1)

3. George area: 8.4 miles southwest on I-90 to the Columbia River; downstream for many miles on the Columbia River, especially on elevated gravel bars.

Agate: pebbles. Petrified wood. (3-7-18-20)

4. Mattawa area: 3 miles east; then north; on slopes of Saddle Mountain (CF).

Opal, wood: logs. (31)

5. Moses Lake; 3 miles west along east side of lake.

Opal, wood: fine; whole logs. Petrified wood: entire forest; limbs and logs. (1)

6. Quincy area.

Opal, common: large brownish nodules; diatomaceous. (6)

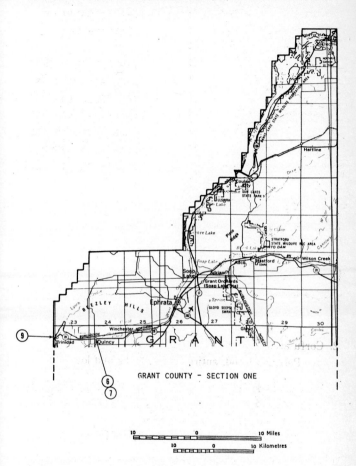

GRANT COUNTY - SECTION ONE

10   0   10 Miles
10   0   10 Kilometres

7. Quincy area: north a few miles.
   Petrified wood: entire forest; limbs and logs. (1)

8. Smyrna area.
   Petrified wood: entire forest; limbs and logs. (1)

9. Trinidad area.
   Opal, wood: logs. Petrified wood. (1)

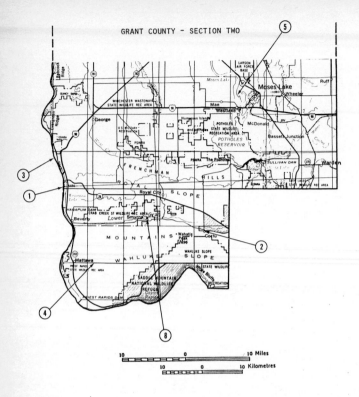

GRANT COUNTY – SECTION TWO

GRAYS HARBOR COUNTY

1. Aberdeen area: stream mouths and beaches on north and south shores of Grays Harbor, all the way to the Pacific Ocean.

    Jasper: flower. (7-20)

2. Grayland area beaches.

    Agate. (20)

3. Malone area: 4 miles northeast.

    Fossils: primarily bivalves; chiefly clams and oysters; all well silicified and agatized. (1-4-7-24)

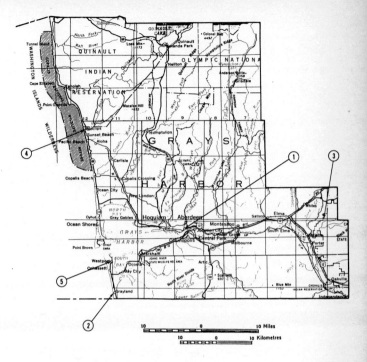

4. Moclips area: on beaches to the north and in stream
   mouths in the vicinity.
   Jasper: orbicular; fine. (20)

5. Westport area ocean beaches, southward to Grayland
   area.
   Agate. (20)

ISLAND COUNTY

1. Whidbey Island beaches.
   Agate. Jade. Petrified wood. (20)

JEFFERSON COUNTY

1. County-wide along Pacific Coast beaches.
   Agate. Petrified wood. (20)

2. Kalaloch area: at Kalaloch Beach.
   Agate. Jasper: orbicular. Petrified wood. (20)

3. Queets area: from beach along the west side of US-101,
   south to the mouth of the Queets River and upstream in
   gravel deposits.
   Agate. Petrified wood. (3-7-20)

4. Uncas area: northeast along the shoreline of the Quimper
   Peninsula to Fort Worden State Park.
   Agate. (20)

JEFFERSON COUNTY
WASHINGTON

DIVIDED INTO TWO
SEPARATE MAPS ON
THE FOLLOWING TWO
PAGES -- EACH MAP
IDENTIFIED BY ITS
NUMBER, AS SHOWN
IN DIAGRAM

JEFFERSON COUNTY
SECTION ONE

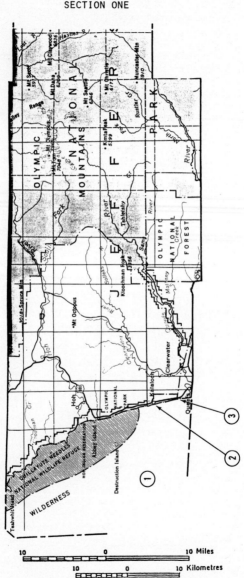

JEFFERSON COUNTY
SECTION TWO

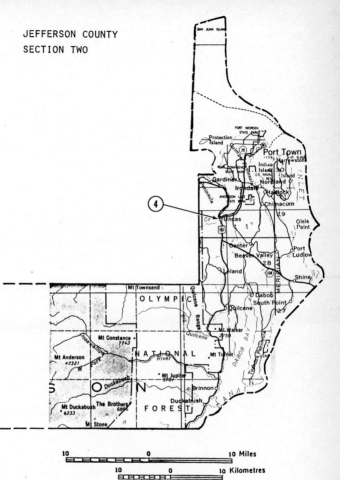

10   0   10 Miles

10   0   10 Kilometres

KING COUNTY

1. Auburn area coal beds.
   Amber (Succinite). (39-coal)

2. Duvall area: 0.6 mile north on SR-203 to fork; bear right and go 5.5 miles northeast, east and south to road east going up hill; follow this 1 mile to fork; bear right to second fork; continue bearing right to gate; walk 1 mile through creek to woods.

   Petrified wood: black. (7-20)

3. Issaquah area: 3 miles south on Hobart Road to left turn on Tiger Road; go 1.5 miles east to a dirt road left; follow latter 0.5 mile to the east side of Fifteenmile Creek; collect at creek and on east side of road for 0.75 mile north.

   Amber (Succinite). Agate: blue. (1-7-13-20)

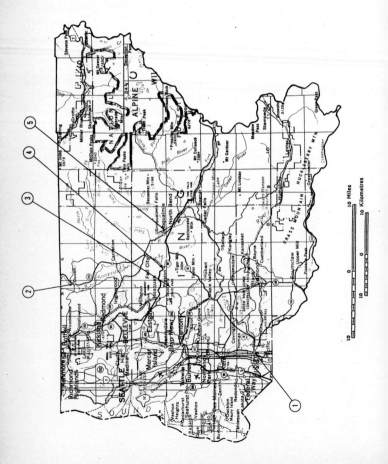

4. Preston area coal beds.

> Amber (Succinite). (39-coal)

5. Snoqualmie area: from Curtiss Campground at Denny Mountain, go 1.5 miles on Denny Creek Trail to gorge.

> Amethyst. Quartz, rock. (1-8-13)

KITTITAS COUNTY

1. Cle Elum area: northeast.

> Agate (Thundereggs): to 6″ diameter; blue. (1-14-ba-salt)

2. Cle Elum area: northeast to steel bridge; cross over and go left to Jack Creek sign; turn right and go 5 miles up mountain to fork; bear left for 0.9 mile; in the cliff and through woods to the right.

> Agate (Thundereggs): to 5″ diameter; blue.

3. Ellensburg area: 13 miles south on US-97 to parking area on right near footbridge; walk across bridge and into hills.

> Agate (Thundereggs): to 6″ diameter; abundant. Chalcedony: geodes to 10″ diameter; abundant. (1-7-8-13-20)

4. Liberty area: upstream along Swauk Creek.

> Agate (Thundereggs): to 6″ diameter; fine. (1-3)

5. Liberty area: at head of Boulder Creek.

> Agate (Thundereggs): to 5″ diameter. (1-7-13-20)

6. Liberty area: at head of Williams Creek.

> Agate (Thundereggs): to 5″ diameter. (1-7-13-20)

7. Liberty area: follow Boulder Creek Road to meadow; then 2 miles on rough road up Robinson Gulch; from there take trail to crest of Crystal Mountain; collect to left of crest.

> Agate. Jasp-agate. Jasper. (1-13)

8. Liberty area: in road cuts for the first 4 miles south on US-97.

   Agate: blue. (1-4-13)

9. Liberty area: north on US-97 to summit; turn right on Lion Rock Road (CR-2107) and then on CR-2101 toward Lookout; 2 miles past slide area: dig on both sides of road.

   Quartz, rock: fluorescent. (6)

10. Liberty area: on Table Mountain.

    Agate (Thundereggs): to 5" diameter. (1-7-13-20)

11. Liberty area: surrounding Lookout Point.

    Petrified wood. (1-3)

12. Liberty area: to Blue Creek Road; follow this for 9 miles; walk up trail to Red Top Lookout; collect in ground just beyond.

    Agate (Thundereggs): to 5" diameter; blue. (1-13)

13. Liberty area: 1 mile northeast; in 2-square-mile area.

    Agate (Thundereggs): to 6" diameter. Chalcedony: geodes to 10" diameter. (1-7-8-13-20)

14. Liberty area: 1 mile west of US-97; on summit of Red Top Mountain.

    Agate (Thundereggs): to 6" diameter. (1)

15. Liberty area: 2.5 miles by trail; at summit of Crystal Mountain.

    Agate (Thundereggs): to 6" diameter. (1)

16. Roslyn area: 4.8 miles northeast to Cumby area: collect upstream on Middle Fork of Teanaway River.

    Agate. Jasp-agate. Jasper. (7-20)

17. Vantage area: just south, in Saddle Mountains; in Crab Creek Canyon.

    Opal, wood: fine; cell structure and annular rings visible. (1)

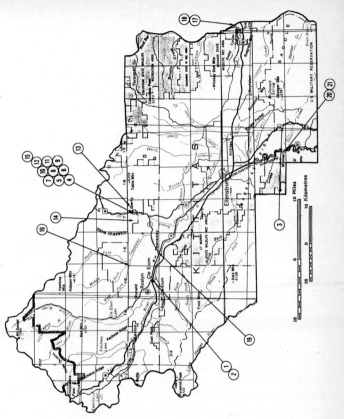

18. Vantage area: 2 miles on Wanapum Road to sign for Petrified Log Trail; pass through gate and stone pillars into Game Department land; continue 2 miles more to another gate.

    Petrified wood. (1-13)

19. Teanaway area: upstream along the Teanaway River.
    Petrified wood. (1-3-7-20)

20. Wymer area: upstream (east) in Squaw Creek.
    Petrified wood. (1-3-7-20)

21. Wymer area: upstream and down along Yakima River.
    Petrified wood. (7-18-20)

KLICKITAT COUNTY

1. Appleton area: 1.2 miles east to crossroad; turn left and go 5.2 miles north to headwaters of Rattlesnake Creek; collect in broad area from here, using this point as center of a roughly circular area 5 miles in diameter.

   Agate. Jasper. Petrified wood. (1-13)

2. Bickleton area: just north and northeast; in Horse Heaven Hills.

   Opal, wood: fine; cell structure and annular rings visible. (1)

3. Glenwood area: northeast in broad area.

   Opal, wood: fine; cell structure and annular rings visible. (1)

4. Goldendale area: 20 miles northeast on US-97; in broad area adjacent to border of Yakima County.

   Agate. Carnelian. Jasper. Petrified wood. (1-3)

5. Goodnoe Hills area: southwest to small bridge; go 0.5 mile up east bank.

   Chert. (1-13-20)

6. Moonax area: 1.5 miles east on SR-14 to Pine Creek; follow road north along the left bank to canyon; then 2 miles to Ford Creek; in east bank of creek.

   Petrified wood. (1-7-13-20)

7. Roosevelt area: northwest from rolling hills on Columbia River.

   Opal, wood: fine; cell structure and annular rings visible. Petrified wood. (1)

8. Warwick area: 2 miles west on SR-8 to bridge over wash; in broad area along and adjacent to Swale Creek.

   Agate. Carnelian. Jasper: green. Petrified wood. (1-3-7-13-20)

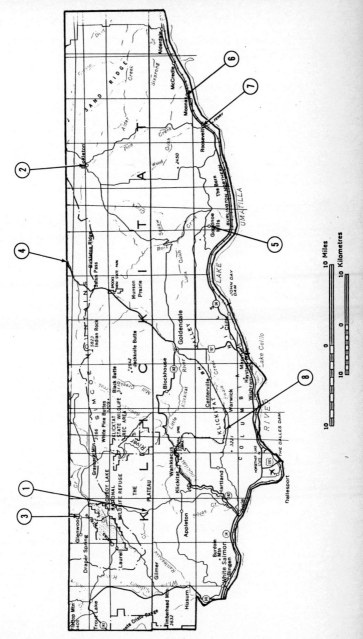

LEWIS COUNTY

1. Adna area: in Chehalis River.
   Agate. (7-18-20)

2. Adna area: west and southwest in road cuts and ditches
   along SR-6 to Pe Ell.
   Agate. (1-4-13)

3. Adna area: 4 miles northwest on Bunker Creek Road; then
   west on Ceres Hill Road; at the McCoy farm (CF).
   Carnelian. Petrified wood. (1-13)

4. Centralia area.
   Carnelian: large. Chalcedony: geodes to 9″ diameter.
   Petrified wood. (7)

5. Centralia area: 4.5 miles east.
   Agate. (1-13)

6. Chehalis area.
   Carnelian: large. Chalcedony: geodes to 9″ diameter.
   Petrified wood. (7)

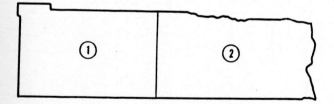

LEWIS COUNTY, WASHINGTON

DIVIDED INTO TWO SEPARATE
MAPS ON THE FOLLOWING TWO
PAGES -- EACH MAP IDENTI-
FIED BY ITS NUMBER, AS
SHOWN IN DIAGRAM ABOVE

LEWIS COUNTY – SECTION ONE

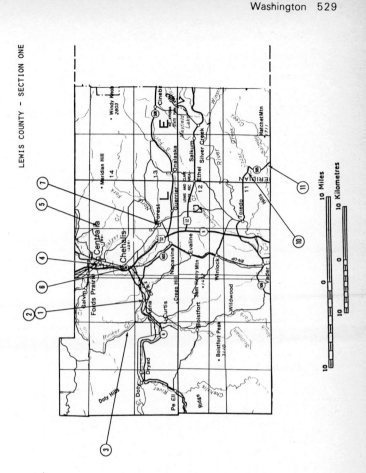

7. Forest area: in North Fork of the Newaukum River.
   Carnelian. (3-7-18-20)

8. Morton area: along Lucas Creek tributary of the Cowlitz
   River.
   Carnelian: large. Chalcedony: geodes to 9″ diameter.
   Petrified wood. (7-20)

9. Morton area: in mercury mines and their dumps.
   Chalcedony: clear; bright red cinnabar inclusions. (2)

10. Toledo area: 6 miles east on SR-505 to bridge over Cedar Creek; collect upstream on both sides.

   Carnelian. (7-13-20)

11. Toledo area: 9.1 miles east on SR-505 to bridge over Salmon Creek.

   Carnelian. (7-13-20)

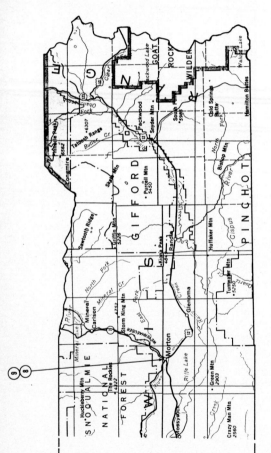
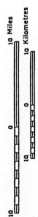

LEWIS COUNTY – SECTION TWO

LINCOLN COUNTY

1. Mondovi area: 1 mile north.
   Opal. (5-vesicular basalt)

2. Mondovi area: 1 mile northwest.
   Opal. (1-5-vesicular basalt -13)

3. Mondovi area: 7.1 miles north to Caps Talc Mine; then 2 miles beyond mine; along south shore of Spokane River in cliffs.
   Agate. (19)

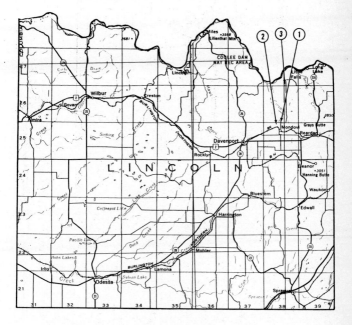

OKANOGAN COUNTY

1. Nespelem area: immediately northwest of town and north-ward upstream to headwaters of Nespelem River.

    Agate, moss. Chalcedony. (1-3-7-13-20)

2. Nespelem area: in sand bars of the Columbia River between the mouths of the Nespelem and Kettle rivers.

    Gold. (17-21)

OKANOGAN COUNTY
WASHINGTON

DIVIDED INTO TWO
SEPARATE MAPS ON
THE FOLLOWING TWO
PAGES -- EACH MAP
IDENTIFIED BY ITS
NUMBER, AS SHOWN
IN DIAGRAM

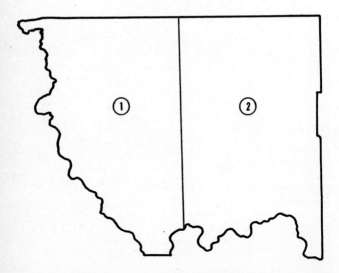

3. Okanogan area: northward up Salmon River to right turn to Happy Hill (CF).

>   Quartz, aventurine. (1-13-29)

4. Oroville area: along shores of saline lakes on Kruger Mountain.

>   Epsomite. Mirabilite. (20)

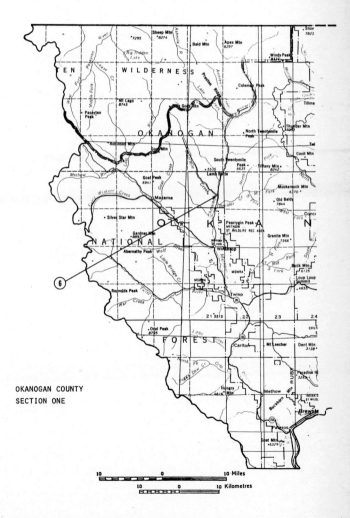

OKANOGAN COUNTY
SECTION ONE

5. Synarep area: from mouth of small creek entering Tunk Creek from north; go upstream along both sides of the tributary creek.

Thulite, pink. (1-5-quartz -7-13-19-20-27-hornblende -29)

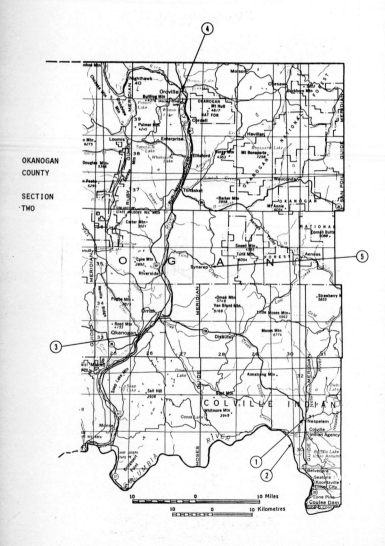

OKANOGAN
COUNTY

SECTION
TWO

6. Winthrop area: 8.1 miles north to Eightmile Creek mouth at Chenwack; upstream on both sides of Eightmile Creek for 10 miles.

   Agate. Chalcedony. (1-3-7-13-20)

PACIFIC COUNTY

1. Lebam area: in Willapa River.

   Fossils, marine: clams and snails; to 1.5″; all well silicified and agatized. (1-3-7-13-20)

2. Raymond area: southwest for 40 miles along SR-6; in virtually every graveled streambed.

   Agate. Carnelian. Chalcedony. (3-7-20)

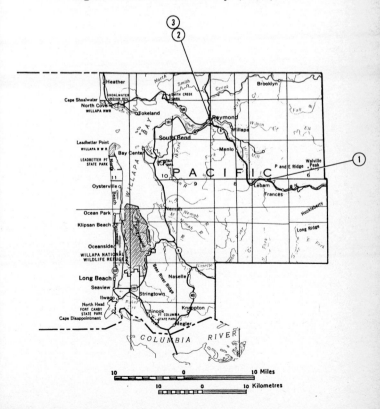

3. Raymond area: southeast to Menlo; along Green's Creek.
   Fossils, marine: ammonites and clams; well silicified.
   (7-14-steep banks -20)

PEND OREILLE COUNTY

1. Metaline Falls area: in the Josephine Mine.
   Smithsonite. (2)

2. Newport area: along Yaquina Bay Park beaches.
   Agate. Jasper. (20)

3. Newport area: 2 miles northwest on the west bank of the
   Pend Oreille River.
   Amethyst. (20)

PIERCE COUNTY

1. Orting area: to Clay City, then 1 mile east; on the Sieg-
   mund ranch.
   Amethyst. (1-13)

SAN JUAN COUNTY

1. San Juan Island in north Puget Sound: on all gravel
   beaches.
   Chalcedony: highest-quality pebbles. (3-20)

SKAGIT COUNTY

1. Anacortes area beaches.
   Jasper. (20)

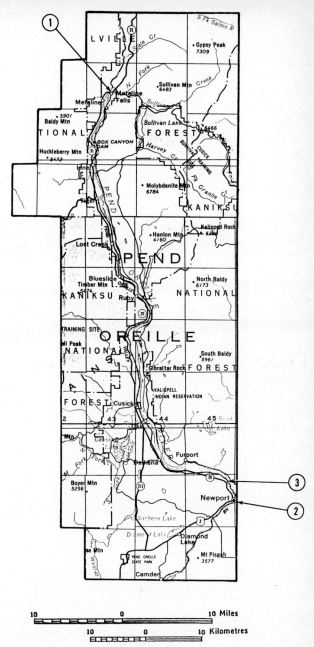

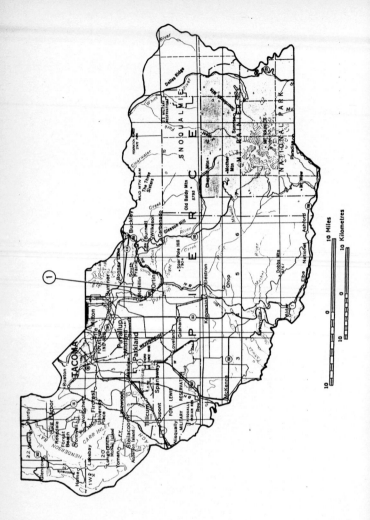

2. Big Lake area: 2 miles south to junction of Walker Valley
   Road; turn left and take Walker Valley Road southeast to
   first side road on right; follow latter southwest 2 miles to
   dig area.

   Agate. Quartz: geodes to 7″ diameter. (1-13-39)

3. Concrete area: south on Dallas Bridge Road toward Darrington; at fork, bear left to Finney Creek Road; turn right and follow Finney Creek on right for 13 miles; turn on Gee Point Road to parking areas; in first creek.

Jade. Serpentine. (7-20)

4. Marblemount area: 14 miles east and southeast on paved road following Cascade River upstream; where pavement ends, collect upstream along north side of Cascade River to confluence of South Fork and North Fork; collect for another mile upstream on both sides of each fork.

Agate. Chalcedony: geodes to 10″ diameter. Jasper. Serpentine. (1-3-7-13-20-29)

5. Prairie area: 2.5 miles east-southeast.

Serpentine. (1-13-29)

SKAGIT COUNTY
WASHINGTON

DIVIDED INTO TWO
SEPARATE MAPS ON
THE FOLLOWING TWO
PAGES -- EACH MAP
IDENTIFIED BY ITS
NUMBER, AS SHOWN
IN DIAGRAM

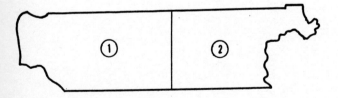

SKAGIT COUNTY
SECTION ONE

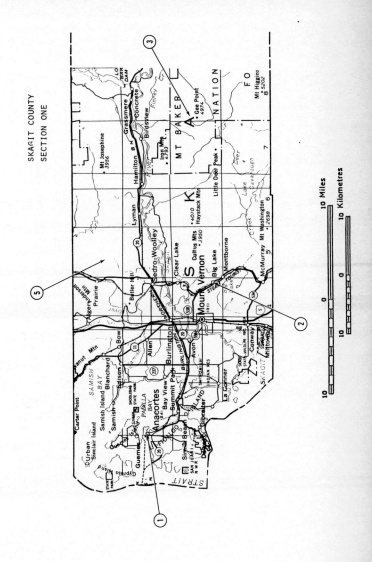

10 Miles

10 Kilometres

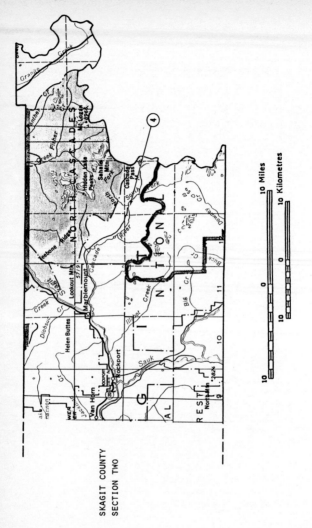

SKAGIT COUNTY
SECTION TWO

SKAMANIA COUNTY

1. Table Mountain area: in northeast corner of county.
   Agate. Carnelian. Jasper. (8-volcanic)

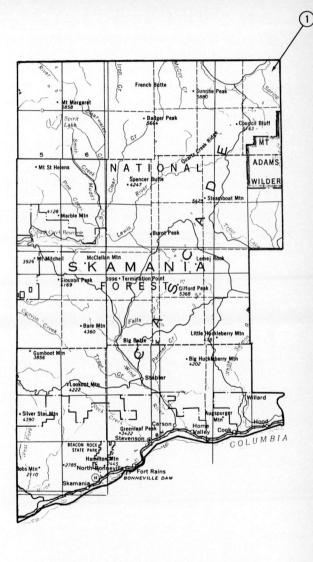

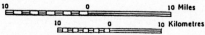

SNOHOMISH COUNTY

1. Oso area: to bridge spanning Deer Creek; collect upstream on both sides of creek.

    Jade: pebbles to boulders. (7-20)

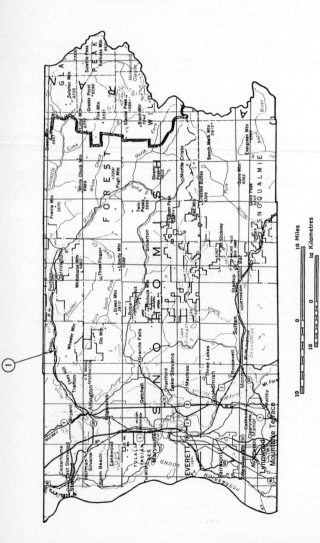

SPOKANE COUNTY

1. Peone area: northeast; in vicinity of Mount Spokane; at the
   Daybreak Mine.

   Autunite: divergent aggregates and crystals; bright
   yellow; well formed. (2)

2. Peone area: northeast; in vicinity of Mount Spokane; at the
   Mudhold Mine.

   Autunite. (2)

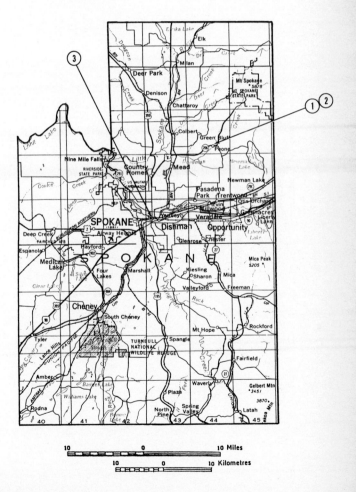

3. Spokane area: at Silver Hill.

Cassiterite. (1-13)

STEVENS COUNTY

1. Cleveland Mine (more precise location unavailable).

Arsenopyrite. Boulangerite: columnar to fibrous masses. Galena. Siderite. (2)

2. Valley area: west on Waitts Lake Road to Carrs Corner; turn left and go 6 miles to quarry.

Calcite (Marble var.): red. (12)

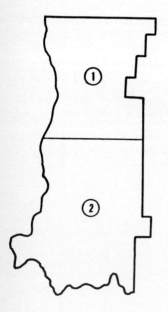

STEVENS COUNTY
WASHINGTON

DIVIDED INTO TWO
SEPARATE MAPS ON
THE FOLLOWING TWO
PAGES -- EACH MAP
IDENTIFIED BY ITS
NUMBER, AS SHOWN
IN DIAGRAM TO LEFT

①

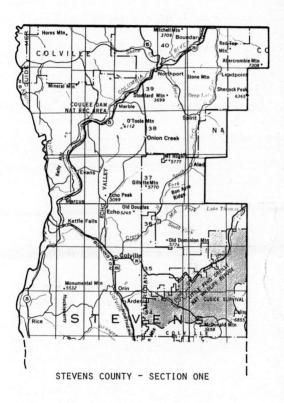

STEVENS COUNTY – SECTION ONE

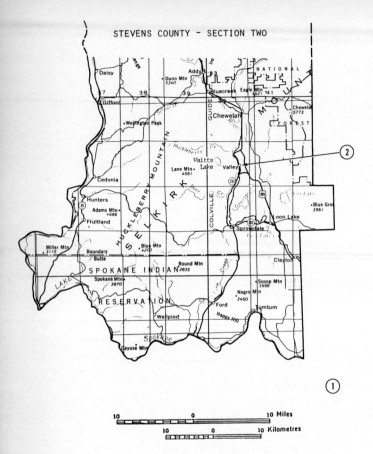

STEVENS COUNTY – SECTION TWO

THURSTON COUNTY

1. Bucoda area: south; at coal mine spoils.
   Agate. Carnelian. (2-7-25-30)

2. Olympia area: in sand around Tono.
   Agate. Petrified wood. (26)

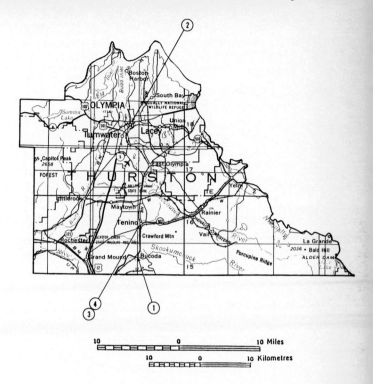

3. Tenino area: south on SR-507 to Skookumchuck Road; then 7 miles east to Johnson Creek Road; then 0.3 mile north to Anderson Ranch (CF).

   Carnelian. Jasper. (1-13)

4. Tenino area: 5.5 miles east on Johnson Creek Road; in ground above creek.

   Carnelian. Jasper. (1-13)

WAHKIAKUM COUNTY

1. Grays River area: at Gorley Ranch (CF).
   Agate. (1-13)

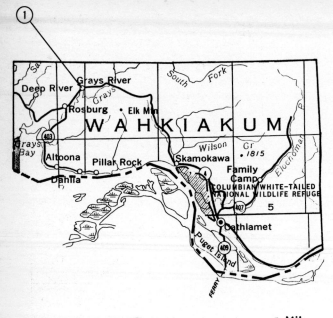

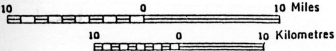

WHITMAN COUNTY

1. Colton area.
      Quartz, smoky. (11)

2. Johnson area: 2.1 miles east; at Bald Butte.
      Corundum. Quartz, smoky. (11)

3. Pullman area: northeast; on hill west of Ringo Station.
      Amethyst. (1-13)

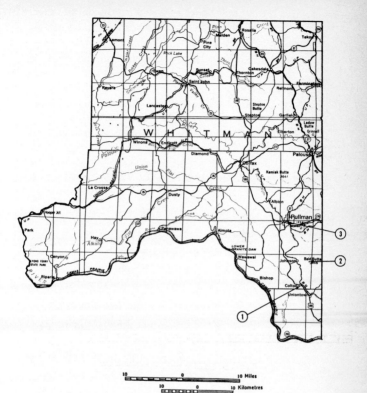

YAKIMA COUNTY

1. Goose Prairie area: upstream and down along north bank
   of Bumping River.

   Agate. Petrified wood. (3-7-20)

2. Mabton area: south to the Horse Heaven Hills; collect on
   the open hills.

   Opal, wood: fine; cell structure and annular rings visible. (1-13)

3. Moxee City; 20 miles east on SR-24 to Goldcreek area; 1 mile northwest; in the Yakima Ridge slopes south of Cairn Hope Peak.

> Opal, wood: fine; cell structure and annular rings visible. (1-13)

4. Pomona area: 13.3 miles east on CR along Yakima Ridge; in broad area south of road in Rattlesnake Hills.

> Opal, wood: fine; cells structure and annular rings visible. (1)

5. Sunnyside area: southeast on US-12 to Vernita Ferry Road; turn north and go 9 miles; at Anderson Ranch on right (CF).

> Petrified wood. (1-13)

6. Yakima area: southwest; on Slide Ranch.

> Petrified wood: entire forest; limbs and logs. (1)

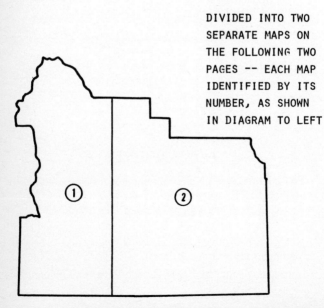

YAKIMA COUNTY
WASHINGTON

DIVIDED INTO TWO
SEPARATE MAPS ON
THE FOLLOWING TWO
PAGES -- EACH MAP
IDENTIFIED BY ITS
NUMBER, AS SHOWN
IN DIAGRAM TO LEFT

YAKIMA COUNTY
SECTION ONE

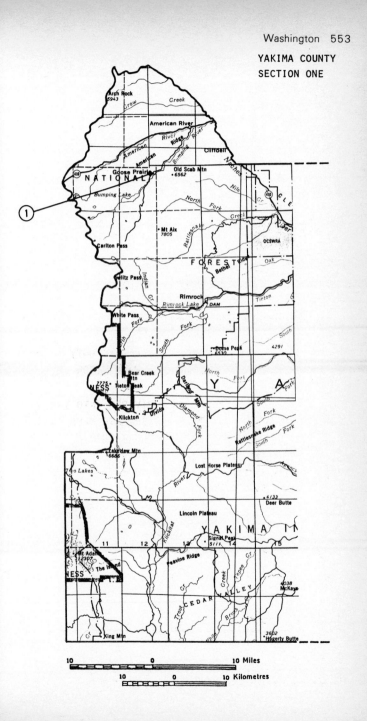

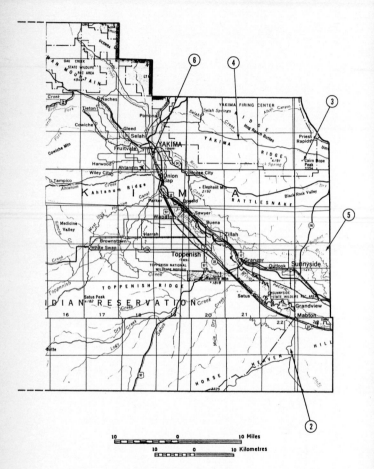

WYOMING

Statewide total of 92 locations in the following counties:

Albany (3)	Hot Springs (2)	Sheridan (2)
Big Horn (1)	Johnson (4)	Sublette (2)
Carbon (5)	Laramie (2)	Sweetwater (10)
Converse (3)	Lincoln (5)	Teton (6)
Crook (3)	Natrona (6)	Uinta (2)
Fremont (21)	Park (3)	Washakie (1)
Goshen (1)	Platte (8)	Weston (2)

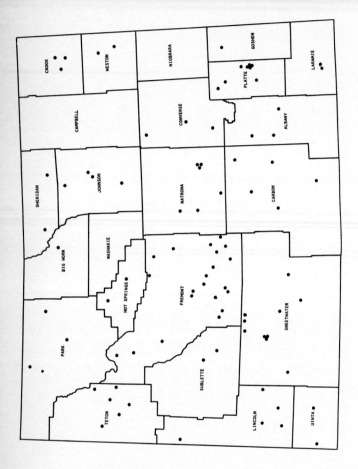

COLLECTING SITES IN WYOMING

(FOR SPECIFIC DETAILS ON THESE
LOCATIONS, SEE THE FOLLOWING
COUNTY MAPS AND RELATED TEXT)

CENTRAL WYOMING REGION

Fine nephrite jade in a very large region roughly 60 miles by 120 miles; within a line running southwest from Lander (Fremont County) through the southeastern portion of Sublette County to Farson (Sweetwater County); then east to Red Desert (Sweetwater County) and northeast from there to Seminoe Dam (Carson County); north-northeast from there to Alcova (Natrona County) and then westward back to the starting point at Lander. The jade is found on the surface and in the soil, often exposed in washes, draws, arroyos, and streambeds.

ALBANY COUNTY

1. Bosler Junction area: 10 miles northeast on SR-34. Jasper. (1-7-13-20)

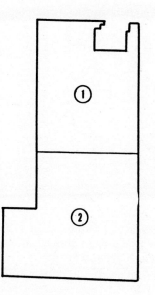

ALBANY COUNTY, WYOMING

DIVIDED INTO TWO
SEPARATE MAPS ON
THE FOLLOWING TWO
PAGES -- EACH MAP
IDENTIFIED BY ITS
NUMBER, AS SHOWN
IN DIAGRAM TO LEFT

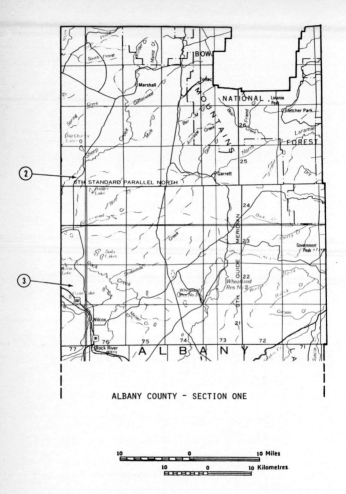

ALBANY COUNTY – SECTION ONE

2. Marshall area: 16 miles south-southwest; 16 miles north-east of Medicine Bow (in Carbon County); in foothills of west slopes of Laramie Mountains.

　　Quartz, cryptocrystalline. (1-13)

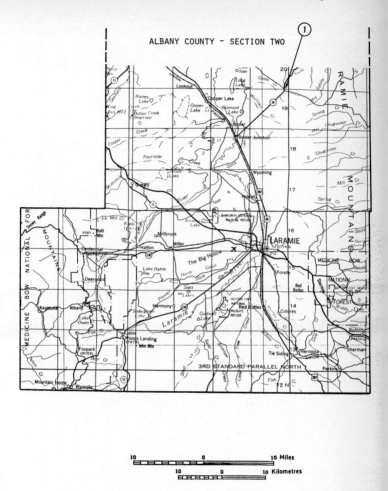

ALBANY COUNTY - SECTION TWO

3. Wilcox area: 5.5 miles northwest; just north of US-30/287
at the famous Como Bluff Dinosaur Graveyard area.
Fossils, bone: dinosaur; silicified. (1-8-13-14)

BIG HORN COUNTY

1. Shell area: 4 miles north; at Elkhorn Ranch.
   Fossils, bone: some silicified. Petrified wood. (1-13)

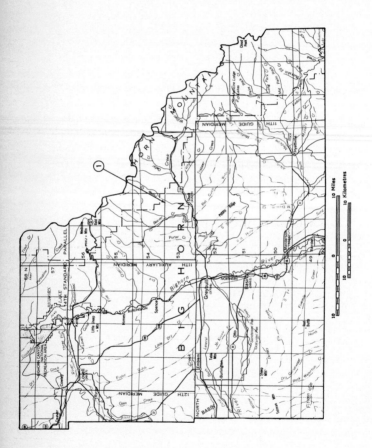

CARBON COUNTY

1. Cow Creek area: collect upstream on Cow Creek to mouth of Calf Creek.

   Agate. Fossils, bone; some silicified. Petrified wood: fine. (1-3-7-20)

2. Kortes Dam area on North Platte River; a few miles below Seminoe Dam.

   Jade, nephrite: black. (3)

3. Medicine Bow area: upstream on both sides of Medicine Bow River; also in tributaries of this river.

   Fossils, bone; some silicified. Petrified wood. (1-8-13)

4. Medicine Bow area: 18 miles north on SR-487 to fork; bear right and continue another 13 miles north; in petrified forest west of road.

   Agate. Petrified wood; limbs and large logs. (1-13)

CARBON COUNTY, WYOMING

DIVIDED INTO THREE
SEPARATE MAPS ON
THE FOLLOWING THREE
PAGES -- EACH MAP
IDENTIFIED BY ITS
NUMBER, AS SHOWN
IN DIAGRAM

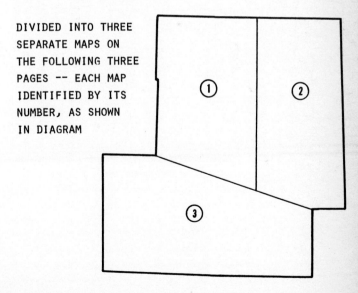

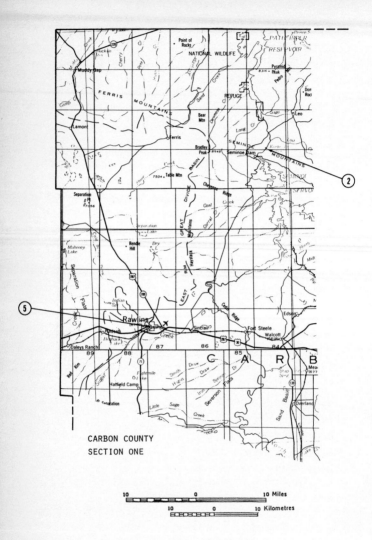

CARBON COUNTY
SECTION ONE

10 ————— 0 ————— 10 Miles

10 ————— 0 ————— 10 Kilometres

5. Rawlins area coal strip mines.
  Amber (Succinite): small amounts. (2-39-coal)

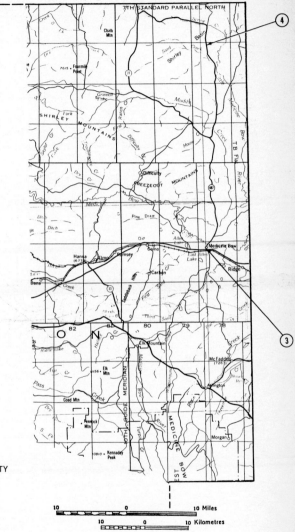

CARBON COUNTY
SECTION TWO

10   0   10 Miles

10   0   10 Kilometres

CARBON COUNTY – SECTION THREE

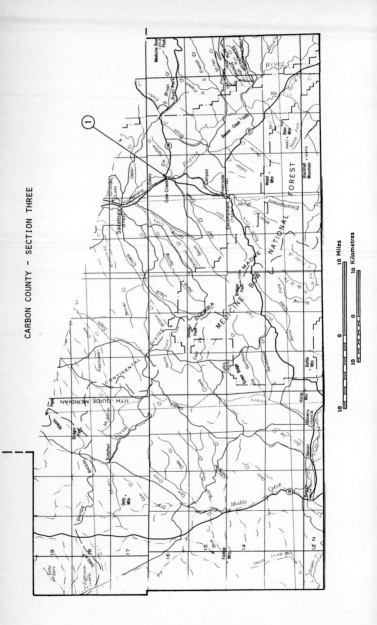

CONVERSE COUNTY

1. Careyhurst area: 2.35 miles southwest to I-25; from mouth of Box Elder Creek in this immediate area, collect upstream (south) along Box Elder Creek.

    Chalcedony: geodes to 14"; many are lined with quartz crystals and/or drusy. (1-3-7-20)

2. Douglas area: 30 miles southwest on SR-91; west of road, on Moss Agate Hill.

    Agate, moss: blue. (1-13)

3. Ross area: collect downstream to mouth of Antelope Creek.

    Agate, moss. (1-13)

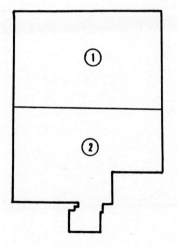

CONVERSE COUNTY, WYOMING

DIVIDED INTO TWO
SEPARATE MAPS
-- EACH MAP
IDENTIFIED BY ITS
NUMBER, AS SHOWN
IN DIAGRAM TO LEFT

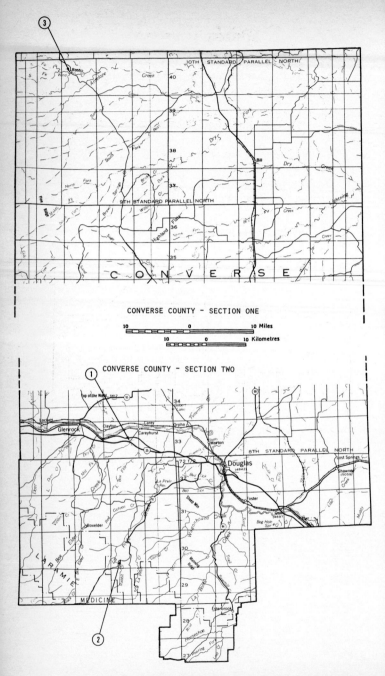

CONVERSE COUNTY - SECTION ONE

10 ⊢⊢⊢⊢⊢⊢⊢⊢⊢ 0 ⊢⊢⊢⊢⊢⊢⊢⊢⊢ 10 Miles

10 ⊢⊢⊢⊢⊢⊢⊢⊢⊢ 0 ⊢⊢⊢⊢⊢⊢⊢⊢⊢ 10 Kilometres

CONVERSE COUNTY - SECTION TWO

CROOK COUNTY

1. Sundance area: just north in numerous Badland exposures.

   Bentonite. (1-13-14-29)

2. Sundance area: 8 miles northwest; at the Copper Prince Mine.

   Chrysocolla. Copper. Gold: occasional. Malachite. (2)

3. Sundance area: 10 miles north; in the Bear Lodge Mountains; at numerous mines in the Black Hills National Forest.

   Fluorite. Gold. (2)

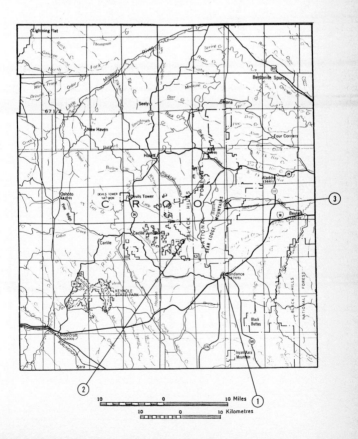

FREMONT COUNTY

1. Atlantic City area: 3 miles northeast; at the Copper Chief Gold Mine.

    Gold. Jade, nephrite. Quartz, rock. (2)

2. Boysen area: 1.1 miles north; just east of Wind River Canyon; virtually on the Hot Springs County border.

    Aquamarine. Beryl. Feldspar. Mica. (16)

3. Dubois area: 9.7 miles east-northeast; at outcrop on East Fork near mouth of Bear Creek.

    Jade, nephrite: olive green. (20-29)

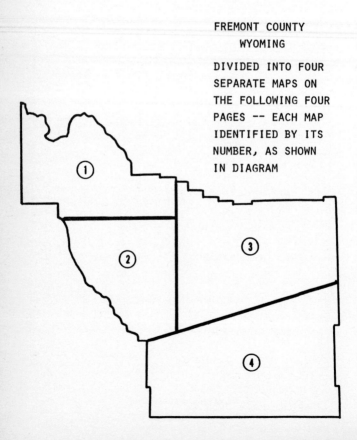

FREMONT COUNTY
WYOMING

DIVIDED INTO FOUR
SEPARATE MAPS ON
THE FOLLOWING FOUR
PAGES -- EACH MAP
IDENTIFIED BY ITS
NUMBER, AS SHOWN
IN DIAGRAM

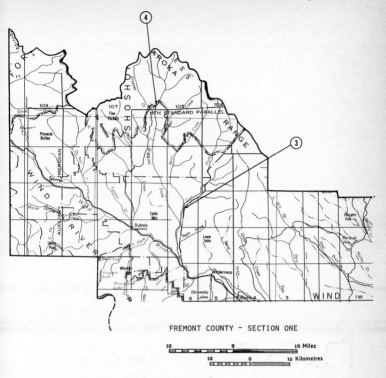

FREMONT COUNTY - SECTION ONE

10   0   10 Miles

10   0   10 Kilometres

4. Dubois area: 20 miles north on paved road to end of pavement; collect upstream (north) in Wiggins Creek (aka Wiggins Fork).

Agate: transparent; dendritic. Petrified wood. (1-7-13-20)

5. Fort Washakie area: 10 miles northwest on US-287; just off road on west side.

Agate, moss. (1-13)

6. Jeffrey City area: 3 miles east on US-287 to a ranch road on left; go through gate and north through ranch yard to flat beyond that.

Agate, moss: fluorescent. (1-13-39)

FREMONT COUNTY - SECTION TWO

7. Jeffrey City area: 3 miles east on US-287 to side road left; go 8 miles south; collect at Green Mountain, especially on the east side of Cottonwood Creek.

    Jade, nephrite: light green. (1-7-13)

8. Jeffrey City area: 8 miles east-northeast on BLM road to the mouth of Sage Hen Creek; in the Sage Hen Creek watershed.

    Agate, moss: fluorescent. (1-13-39)

9. Jeffrey City area: 10.9 miles south to fork; collect eastward in foothills of Green Mountain.

    Jade, nephrite. (1-13-29)

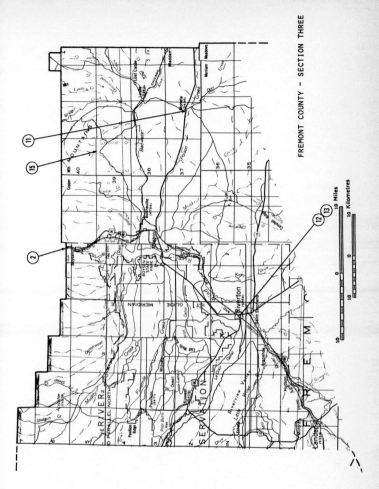

10. Jeffrey City area: 13.75 miles east on SR-789; in the area where Fremont, Natrona, and Carbon counties meet.

Agate. Jade, nephrite: pebble-size to boulders. (1-3-39)

11. Moneta area: at Marston Lake.

Jade, nephrite. (1-7-13-20)

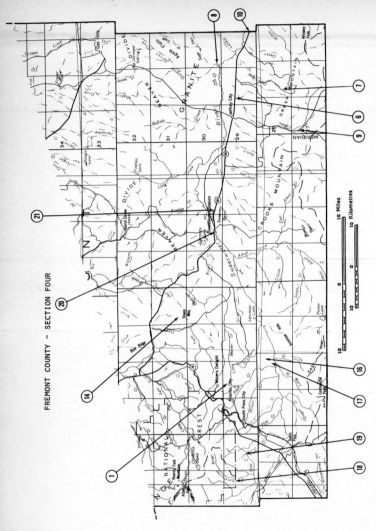

12. Riverton area: on a principal dirt road leading west; park
where powerlines cross overhead; collect in washes northeast.

Agate: nodules to 7″ diameter. (1-13-39)

13. Riverton area: southwest; along the Wind River and its tributaries.

    Agate. (1-3-7-18-20-39)

14. Sheep Mountain area: along the divide between Hall Creek and Twin Creek.

    Petrified wood. (1-13)

15. Shoshoni area: 15 miles northeast; in dikes in Bridger Mountains.

    Aquamarine. (16)

16. South Pass City area: at Split Rock area; on the Sweet-water Divide.

    Ruby. (27)

17. South Pass City area: near Split Rock; just north of the mouth of Long Creek at the Sweetwater River, where a series of low hills are cut by dikes.

    Jade, nephrite: boulders. (16-hornblende/augite)

18. South Pass City area: 1.8 miles due north to SR-28 to northeast-facing sign for Bridger Wilderness; go south-west on SR-28 and take next dirt road left; stay on the main part of that road for 9 miles, heading toward two flat-topped buttes; at sharp turn in road, park and collect east-ward and southward.

    Agate. Petrified algae. Petrified wood. (1-13-39)

19. South Pass City area: 4.8 miles due west to Gold Creek (2.6 miles west of SR-28); collect in Gold Creek downstream to its mouth at the East Sweetwater River and upstream to its headwaters.

    Jade, nephrite. (1-3)

20. Sweetwater Station area: in valleys, washes, plateaus, and most ground surfaces of Townships 30 and 31 North and Ranges 89, 90, and 91 West.

    Quartz, cryptocrystalline (aka Sweetwater agate): nodules to 2″ diameter; many dendritic. (1-3-39)

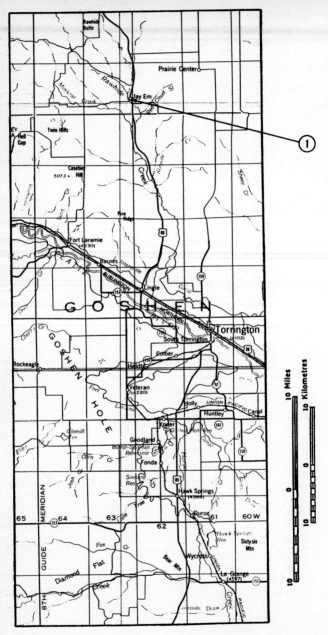

21. Sweetwater Station area: 3.2 miles east on US-287; at the Abernathy Claim.

> Jade, nephrite. Quartz, rock. (2)

GOSHEN COUNTY

1. Jay Em area quarry.

> Calcite (Onyx var.). (12)

HOT SPRINGS COUNTY

1. Grass Creek area: upstream on both sides of Grass Creek to its headwaters.

> Agate. Petrified wood. (1-3-7-13-18-20)

2. Thermopolis area.

> Gypsum (Alabaster var.). (1-3-21)

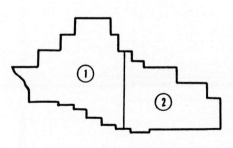

HOT SPRINGS COUNTY
WYOMING

DIVIDED INTO TWO
SEPARATE MAPS
-- EACH MAP
IDENTIFIED BY ITS
NUMBER, AS SHOWN
IN DIAGRAM TO LEFT

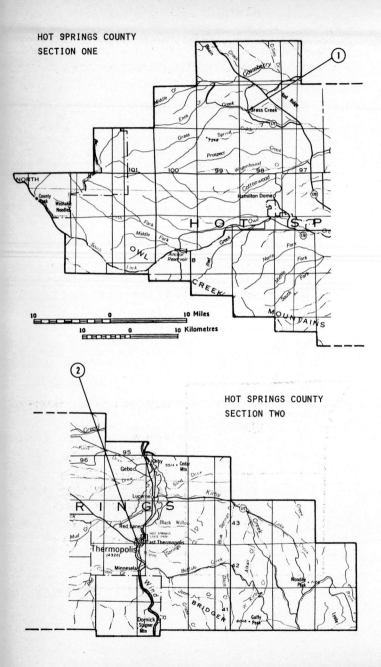

HOT SPRINGS COUNTY
SECTION ONE

HOT SPRINGS COUNTY
SECTION TWO

JOHNSON COUNTY

1. Buffalo area: 7 miles south on SR-196 to side road left; follow latter 11.75 miles east and southeast to bridge over Crazy Woman Creek; collect upstream and down along both sides of Crazy Woman Creek.

    Petrified wood. (3-7-20)

2. Buffalo area: 12 miles northeast on US-16; along both sides of road.

    Agate. Chalcedony. Petrified wood. (1)

3. Buffalo area: 15.5 miles southeast; east side of Big Horn Mountains; near Crazy Woman Creek; in the Crazy Woman Petrified Forest.

    Petrified wood: huge logs. (1-13)

4. Kaycee area: upstream and down on both sides of the Middle Fork of the Powder River.

    Agate. Petrified wood. (1-3-7-13-20)

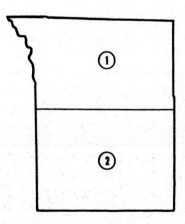

JOHNSON COUNTY, WYOMING

DIVIDED INTO TWO
SEPARATE MAPS ON
THE FOLLOWING TWO
PAGES -- EACH MAP
IDENTIFIED BY ITS
NUMBER, AS SHOWN
IN DIAGRAM ON LEFT

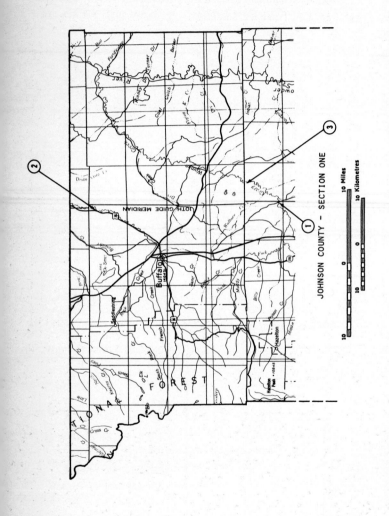

JOHNSON COUNTY – SECTION ONE

JOHNSON COUNTY – SECTION TWO

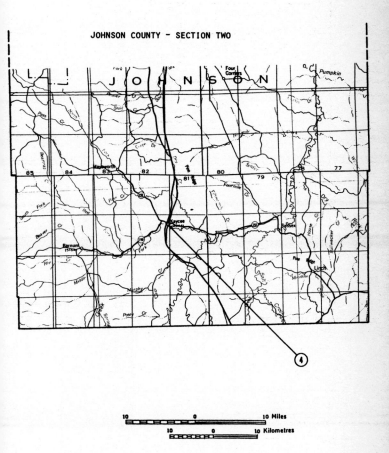

10   0   10 Miles

10   0   10 Kilometres

LARAMIE COUNTY

1. Cheyenne area: in Cheyenne Pass.
   Quartz, rock. (1-13)

2. Cheyenne area: just west.
   Chalcedony. (3)

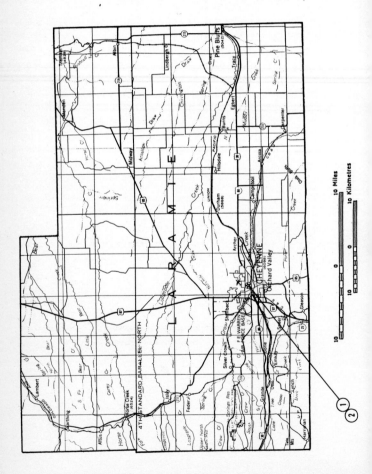

LINCOLN COUNTY

1. Alpine area: upstream on both sides of the Snake River
   and on the adjacent north slopes of Bradley Mountain.

   > Agate. Chalcedony. Jasper. Petrified wood. (1-3-7-13-
   > 20)

2. Diamondville area: 6 miles south on US-189 to gravel side
   road right with sign for Fish Fossils; 6.5 miles on latter
   road to Warfield Springs Fossil Quarries (CF). (Advance
   reservations are recommended: Box 316, Hwy 89, Thane,
   WY 83127; phone 307/883-2445)

   > Fossils: superb fossil fish and other vertebrate and
   > invertebrate animals. (12)

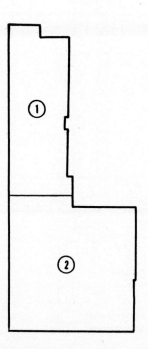

LINCOLN COUNTY
WYOMING

DIVIDED INTO TWO
SEPARATE MAPS ON
THE FOLLOWING TWO
PAGES -- EACH MAP
IDENTIFIED BY ITS
NUMBER, AS SHOWN
IN DIAGRAM TO LEFT

3. **Fontenelle area:** encompassing the entire eastern portion of Lincoln County and continuing into adjacent Sweetwater County; especially in the Badlands area along the Green River and Hams Fork.

    Agate. Chalcedony. Jasper. Petrified algae. Petrified wood. Quartz, cryptocrystalline. (1-2-7-13-20)

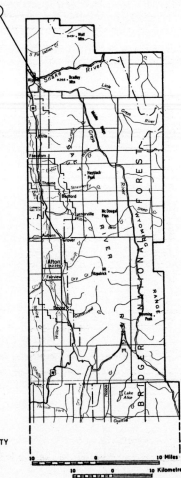

LINCOLN COUNTY
SECTION ONE

4. Kemmerer area: 9 miles north on SR-233 to dim trail west; follow this a short distance to Hams Fork; collect upstream and down on both sides.

   Fossils, vertebrate: fish; very fine; whole and fragments in shale splittings. (1-13-20-28)

5. Opal area: surrounding town.

   Petrified algae. (1-13)

LINCOLN COUNTY - SECTION TWO

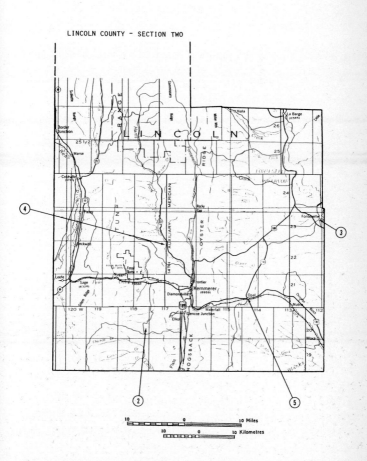

NATRONA COUNTY

1. Casper area.
   Gypsum (Alabaster var.). (1-3-21)

2. Casper area: along northwest bank of the North Platte River, all the way downstream (southwest) to Glenrock (in Converse County).
   Petrified wood. (20)

3. Casper area: west; in broad area lying between the South Fork of Casper Creek and Poison Spider Creek.
   Carnelian. (1-13)

4. Ervay area: 11.75 miles southeast to where the road runs along the north side of Poison Spider Creek; in ground between road and creek and along north shore of creek.
   Jasper. Quartz, cryptocrystalline. (1-3-7-13-20)

5. Hells Half Acre area: 2 miles west on US-20/26 to bridge over tributary of South Fork; upstream on both sides to headwaters.
   Jasper. Quartz, cryptocrystalline. (1-7-20)

6. Pathfinder Dam area: also along shores (especially southwest) of Pathfinder Reservoir.
   Jade, nephrite. (20-30)

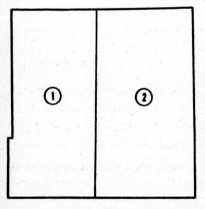

NATRONA COUNTY, WYOMIN

DIVIDED INTO TWO
SEPARATE MAPS ON
THE FOLLOWING TWO
PAGES -- EACH MAP
IDENTIFIED BY ITS
NUMBER, AS SHOWN
IN DIAGRAM TO LEFT

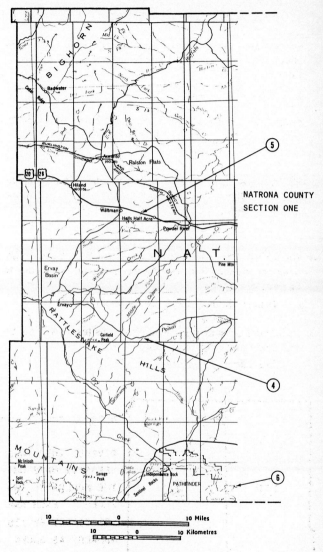

NATRONA COUNTY
SECTION ONE

10 Miles

10 Kilometres

NATRONA COUNTY
SECTION TWO

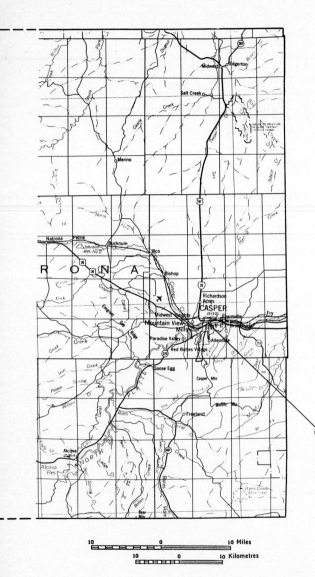

10   0   10 Miles

10   0   10 Kilometres

PARK COUNTY

1. Clarks Fork area: 6 miles southeast of Cooke City, Montana, on US-212; downstream from there to mouth of principal north-flowing drainage from Windy Mountain.

    Agate. Carnelian. Jasper. Petrified wood: fine. (1-3-7-13-18-20)

2. Wapiti area streambeds.

    Agate, moss. (7-20)

3. Willwood area: downstream along south shore of Shoshone River.

    Agate. Jasper. Petrified wood. (3-7-20)

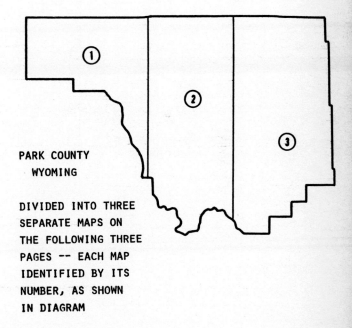

PARK COUNTY
WYOMING

DIVIDED INTO THREE
SEPARATE MAPS ON
THE FOLLOWING THREE
PAGES -- EACH MAP
IDENTIFIED BY ITS
NUMBER, AS SHOWN
IN DIAGRAM

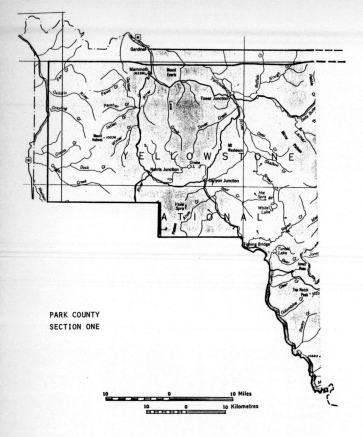

PARK COUNTY
SECTION ONE

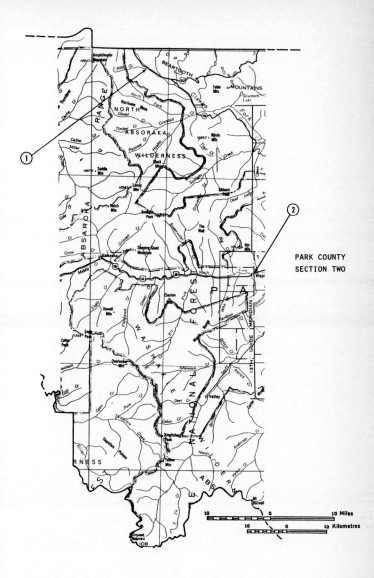

PARK COUNTY
SECTION TWO

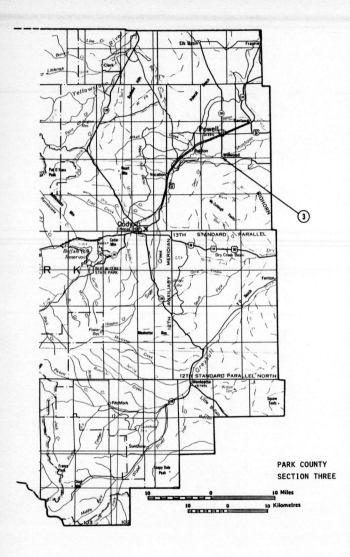

PARK COUNTY
SECTION THREE

PLATTE COUNTY

1. Cassa area: upstream and down on west side of North Platte River; in hills alongside the river.

    Agate. Chalcedony. Petrified wood. (1-13)

2. Glendo area: to northeast in hills flanking the Glendo Reservoir.

    Chalcedony. (1-13)

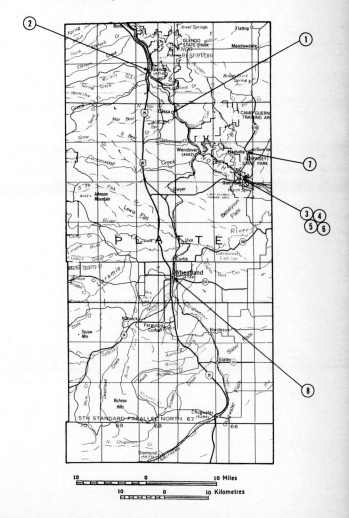

3. Guernsey area: along shores of the Guernsey Reservoir; especially the west shore and also the northwest shore near caves.

   Agate. Jasper. Quartzite. Youngite. (20)

4. Guernsey area: just north.

   Calcite (Onyx var.). (39)

5. Guernsey area: collect northward along west side of SR-270 to Hartville.

   Agate. (1-13)

6. Guernsey area: north on SR-270; at quarry.

   Agate. (12)

7. Hartville area: at the Wilde & Deercorn Mine.

   Agate, moss; dendritic; up to 1000 lb; translucent; bluish opacity. (2-19)

8. Wheatland area: collect to south and west along both sides of SR-34.

   Labradorite. Moonstone. (1-13)

SHERIDAN COUNTY

1. Burgess Junction area: in dikes running just north of the Burgess Ranger Station.

   Leopardite (granite). (16)

2. Ulm area: collect downstream (east-southeast) along Lone Tree Creek for 7 miles to the mouth of Pinhead Creek; pay particular attention to the mouths of washes entering Lone Tree Creek, especially from the south.

   Agate, moss. Jasper. Petrified wood: quality. (1-13)

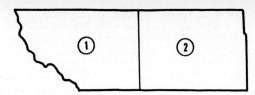

SHERIDAN COUNTY
WYOMING

DIVIDED INTO TWO
SEPARATE MAPS
-- EACH MAP
IDENTIFIED BY ITS
NUMBER, AS SHOWN
IN DIAGRAM

SHERIDAN COUNTY
SECTION ONE

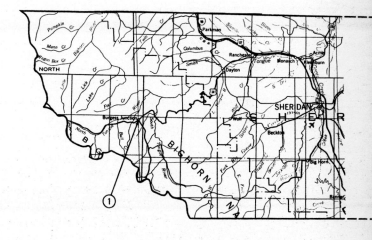

SHERIDAN COUNTY
SECTION TWO

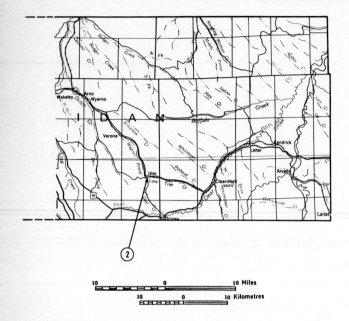

SUBLETTE COUNTY

1. Big Sandy area: 7.4 miles southeast to the bridge over the Big Sandy River; collect downstream all the way to Big Sandy Reservoir and upstream to west slopes of the Wind River Range.

    Petrified wood: fine. (1)

2. Pinedale area: 5.8 miles north-northeast to trail heading right; follow this 0.8 mile to Half Moon Lake; collect along shoreline.

    Agate. Petrified wood: fine. (1-13-20)

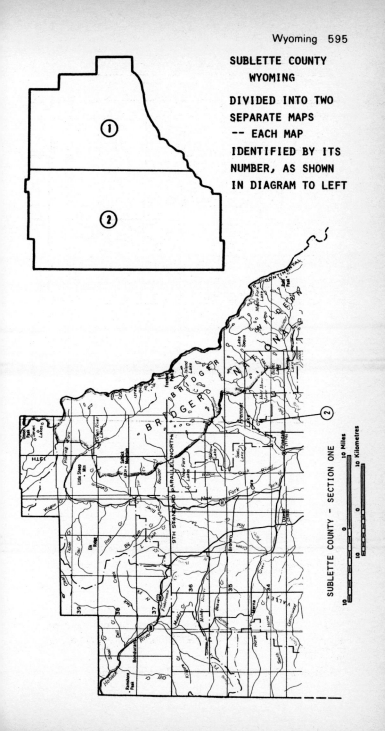

SUBLETTE COUNTY
WYOMING

DIVIDED INTO TWO
SEPARATE MAPS
-- EACH MAP
IDENTIFIED BY ITS
NUMBER, AS SHOWN
IN DIAGRAM TO LEFT

SUBLETTE COUNTY - SECTION ONE

SUBLETTE COUNTY - SECTION TWO

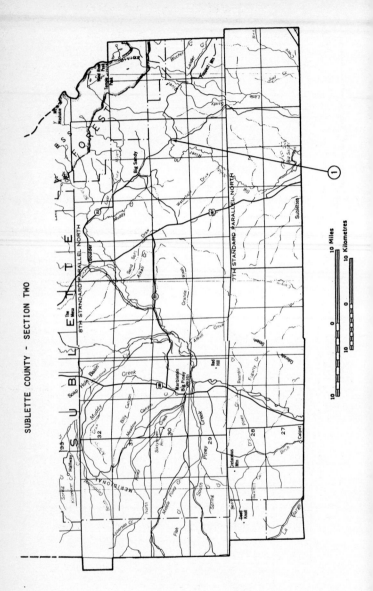

SWEETWATER COUNTY

1. Eden area: collect eastward for 40 miles in desert country to Red Desert.

    Petrified wood: black; coated with dense ivory-colored silica. (1-13)

2. Eden area: collect northeastward for many miles.

    Agate. Chalcedony. Petrified wood. Quartz, cryptocrystalline. (1-3-13)

3. Eden area: broad collecting area some 25 miles wide, south-southeast; 20 miles east of Green River and running southwest for about 65 miles to Fort Bridger (in Uinta County).

    Chalcedony. Jasper. Petrified wood. Quartz, cryptocrystalline. (1)

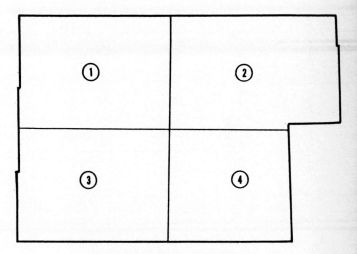

SWEETWATER COUNTY, WYOMING

DIVIDED INTO FOUR SEPARATE
MAPS ON THE FOLLOWING FOUR
PAGES -- EACH MAP IDENTI-
FIED BY ITS NUMBER, AS
SHOWN IN DIAGRAM ABOVE

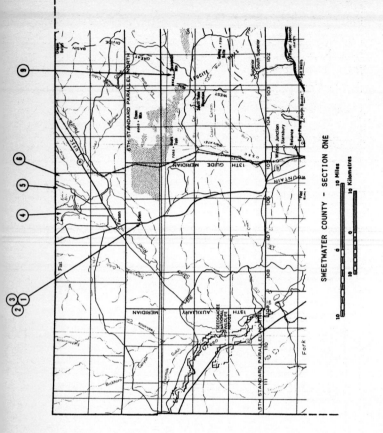

4. Farson area: northeast up the Big Sandy River and its tributaries.

Petrified palm: to 9″ diameter. Petrified wood: fine; chips, branches, logs. (1-6-13-39)

5. Farson area: 7.8 miles north on US-187; turn right on dirt side road and go 2.75 miles east-northeast to Big Sandy Reservoir Dam; cross over dam and continue just over 1 mile to fork; bear right (straight); collect along the road on both sides for several miles.

Petrified wood: whole limb casts. (1-13-39)

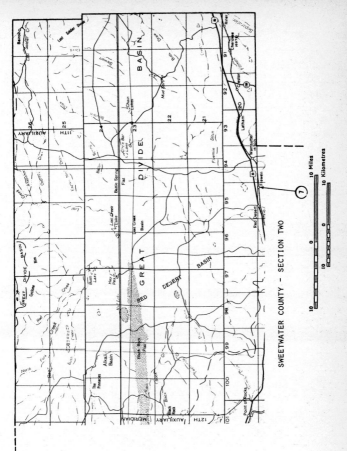

SWEETWATER COUNTY - SECTION TWO

6. Farson area: 9 miles north on US-187; turn right on dirt
   side road and go 2.75 miles to Big Sandy Reservoir Dam;
   cross over dam and continue just over 1 mile to fork; bear
   right (straight) for 1.5 miles to another fork; continue
   straight 1.5 miles to cross-trail; continue straight beyond
   cross-trail for 2.3 miles to center of collecting area: best
   collecting northward from this point; probing and digging
   is most effective for best specimens.

   Petrified bamboo (cane) sections to 3" diameter and 7"
   long; some with solid agate centers; some hollowed
   and lined with agate and/or crystals. (1-6-13)

SWEETWATER COUNTY - SECTION THREE

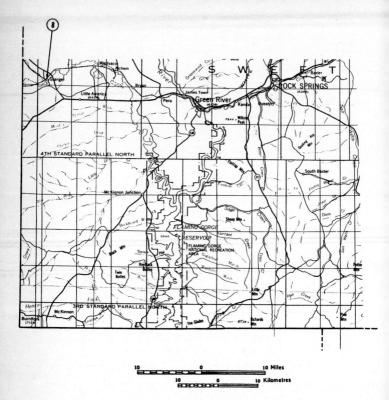

7. Frewen area: just south; on edge of the Delaney Rim.
   Fossils, marine: turritella shell; silicified (aka Turritella agate); very abundant. (1-13-42)

8. Granger area: on ledge atop buttes along US-30 and southwest to Blacks Fork.
   Fossils: algae; silicified. Fossils, marine: turritella shell; silicified (aka Turritella agate). (1-13-42)

9. Superior area: 12.6 miles north; on slopes of Steamboat Mountain and in the Leucite Hills.
   Agate. Chalcedony. (1-13)

SWEETWATER COUNTY - SECTION FOUR

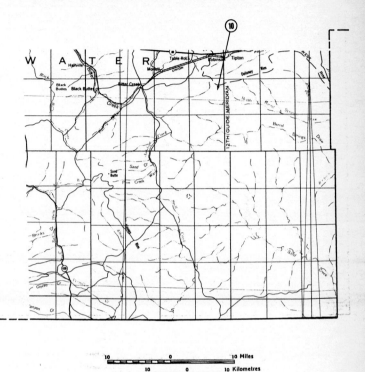

10. Tipton area: 1.8 miles south to fork; bear right and go another 4 miles to within sight of the Cathedral Bluffs on Delaney Rim; take road up bluffs to crest and turn left a short distance.

Agate. Fossils: algae; brown; silicified. Fossils, marine; turritella shell (aka Turritella agate). (1-13)

TETON COUNTY

1. Absaroka Mountains area: on southwest slopes in area of Atlantic Creek.

    Opal, wood. (1-3-13)

2. Moran area: north on US-89/287 to 6 miles south of Yellowstone National Park entrance; then 4.25 miles west; collect along headwaters of Berry Creek, especially below east slope of Survey Peak.

    Amphibolite. Chrysotile. (1-7-13-20)

3. Moran area: 2.75 miles east to side road left; follow this and/or US Forest Service roads upstream along Buffalo Fork of the Snake River; at series of mines there.

    Azurite. Chalcocite. Chalcopyrite. Copper. Malachite. Pyrite. Quartz, rock. (2)

4. Moran area: 3 miles southwest; in Black Rock Meadows.

    Bentonite. (1-13-39)

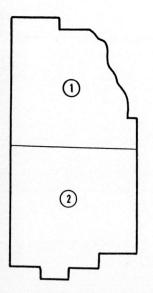

TETON COUNTY, WYOMING

DIVIDED INTO TWO
SEPARATE MAPS ON
THE FOLLOWING TWO
PAGES -- EACH MAP
IDENTIFIED BY ITS
NUMBER, AS SHOWN
IN DIAGRAM TO LEFT

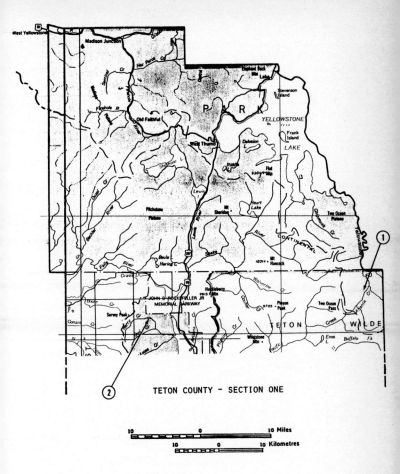

TETON COUNTY — SECTION ONE

5. Moran area: 10 miles south; in Teton Canyon.
   Bentonite. (1-13-14-29)

6. Moran area: 19.5 miles east on US-26/287 to crest of Tog-
   wotee Pass; at elevation 9658'.
   Bentonite. (1-13-14-29)

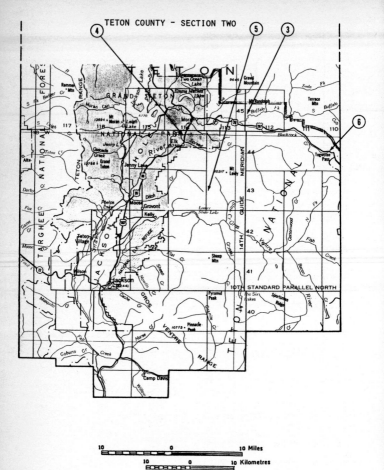

TETON COUNTY - SECTION TWO

UINTA COUNTY

1. Evanston area: 6 miles south; just west of Myers Bridge.
   Bentonite. (1-13-14-29)

2. Fort Bridger area.
   Agate, common. Agate, turritella. Fossils, algae: silicified. Jasper. (1-3-7-13-20)

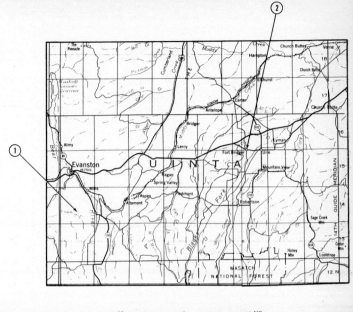

WASHAKIE COUNTY

1. Worland area: just west of town.
   Chalcedony. (1-13)

WESTON COUNTY

1. Newcastle area: upstream and down along Stockade Beaver Creek.
   Gypsum: good crystals. (1-13-20)

2. Newcastle area: 10 miles north; 2 miles southeast of Mount Pisgah; in valley.
   Gypsum: good crystals. (1-13-39)

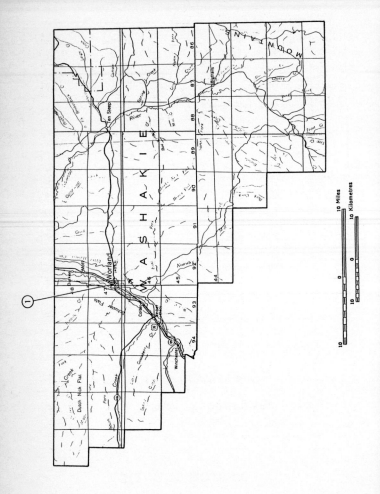

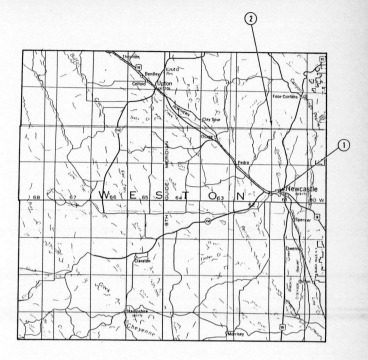

10   0   10 Miles

10   0   10 Kilometres

GLOSSARY

Accessory minerals   Minerals usually present in igneous rocks, but only in minor abundance, their presence not being essential to the size, shape, or definition of the rock.

Accidental minerals   Minerals found (sometimes in large quantities) in a particular rock type where they are not normally associated.

Acicular   Describing a crystal that is considerably elongated, often to the point of being needle-shaped.

Adamantine   Exhibiting a bright luster, as the diamond does.

Adit   A mine entryway from outside; usually horizontal or only slightly inclined.

Aggregate   A mass of rocks or minerals naturally cemented together.

Alluvial   Pertaining to sand, silt, clay, and gravel deposits laid down by running water, most often in areas where the water flow no longer exists or is periodic.

Ameboid   Featuring a coating of minute lobate crystals imparting a distinct texture.

Amorphous   Describing a mineral or rock mass lacking crystal structure; literally, a formless mineral mass (e.g., opal).

Amygdule   A usually round cavity in volcanic rock formed by cooling of the fluid rock around a gas bubble; the cavity is very often lined or filled with secondary minerals, often crystallized.

Anhedral crystals   Crystals that, due to varying causes, do not exhibit their usual crystal form.

Argillaceous   Largely comprised of (or containing) clay (e.g., some shales and slates).

Arroyo   The channel, usually with steep banks, cut by a creek; its bed is ordinarily dry or has only rare and intermittent flowage.

Ash   Very tiny volcanic cinders that have not become solidly cemented.

Asparagus stone A chartreuse (yellow-green) apatite.

Asterism The property of reflecting light in starlike patterns of 4, 6, or 12 rays (and, rarely, 24 rays). Sapphire and ruby are examples. (See also *stellate.*)

Badlands A region, usually of desert characteristic, essentially barren of vegetation, in which erosion (by wind and, less often, water) has carved the soft sedimentary deposits into rather incredible pinnacles, formations, and gullies (e.g., the South Dakota Badlands).

Banded Describing a form of agate (and often other minerals) in which a pattern of arced (rather than angular) stripes occur, usually in concentric manner and either of different colors or gradations of the same color. (See also *fortification.*)

Bar A deposit of gravel, sand, or mud in streams or rivers where currents are less swift, as on the inside of bends and at the mouths of streams.

Basalt A very dense dark rock formed by solidified lava and mainly comprised of plagioclase feldspar and pyroxene; often contains crystal-producing gas cavities; frequently found in flowage columns as the result of contraction during cooling, creating a tree-trunk appearance (e.g., Wyoming's Devils Tower).

Batholith A vast mass of igneous rock (cooled molten rock) that formed deep underground.

Bed Rocks (usually of a single type) laid down in strata forming roughly parallel layers.

Bladed A crystal structure that is thin, flat, and elongated.

Bolson A relatively level, often roughly circular valley draining to a central basin or pan, surrounded by high terrain. (See also *playa.*)

Botryoidal Formed into masses of small symmetrically spherical bulges; often in grapelike clusters.

Breccia A mass of jagged or angular mineral fragments naturally cemented together.

Brittle Easy to break.

Butte A steep-sided and usually isolated large hill or small mountain.

Calcareous Composed of (or containing) calcite or calcium carbonate.

Capillary Extremely elongated hairlike crystal form.

Carat A gemstone weight unit equivalent to 200 milligrams (0.20 gram); there are 141.75 carats to the ounce (avoirdupois).

Cast A mineral that takes the shape of a cavity in rock left by the disintegration of plant or animal matter.

Cat's-eye A translucent stone whose inner fibers reflect light in a sharp streak across the surface.

Cattle-guard A barrier for cattle, constructed across an opening in a fence or at a gate through the parallel placement of slightly separated pipes (usually over a ditch or trench) upon which cattle cannot secure safe footing.

Cephalopod An ancient fossil invertebrate, cylindrical in shape and segmented (often internally silicified) and distantly related to the modern squid.

Cerro Hill or (sometimes) mountain.

Chatoyant Having the property of reflecting light in an undulating or moving streak across the surface.

Cinder cone The naturally expelled debris, usually in a conical pile, surrounding the vent of a volcano.

Claim A piece of land legally filed upon and held for the purpose of mineral prospecting and exploitation; the maximum size a lode claim can be in the United States is 600′ × 1500′.

Clast A rock fragment or mineral fragment, the fragmentation having occurred naturally, through volcanic explosion, earthquake, or meteoritic impact or unnaturally during transportation.

Clay Exceptionally fine-grained earthy substance derived from the weathering of granite and other aluminous rock.

Cleavage The smooth plane surface along which a mineral splits or parts.

Colloform Having a smoothly rounded kidney shape. (see *reniform*.)

Color zone The unblended layering, segregation, or striping of different colors within a crystal (e.g., tourmaline).

Columnar Describing a crystal that has the shape of a column.

Conchoidal Describing a concave fracture in a glassy mineral, the sort of fracture made by a pellet or BB striking plate glass; obsidian is markedly conchoidal in its fracturing.

Concretion A generally spherical formation (often so precise as to appear artificial) caused by the natural cementing together of particles of gypsum, silica, and other minerals in sedimentary rock; easily freed from matrix material; regularly weathers free.

Conglomerate Also called *puddingstone;* rounded pebbles of essentially the same size that have been naturally cemented together.

Contact The zone where differing rock types meet.

Coprolite The fossilized excrement of reptiles, birds, and mammals.

Core, pegmatite The central mass of rock (often a bean-shaped or disk-shaped mass of quartz) found within a pegmatite body.

Coulee In the northwestern states, a very steeply walled river valley or dry riverbed.

Country rock The rock mass that surrounds ore bodies, veins, dikes, and the like.

Cove A relatively small, level area intruding upon a mountainous area.

Crater The natural depression marking the vent of a volcano or impact site of a meteorite.

Crystal A mineral mass, solid and having a regular geometric shape, with smooth flat planes (crystal faces or, erroneously, facets).

Craze The network pattern of fine cracks that develop on the outer surface of some gemstones.

Crinoid An ancient invertebrate fossil animal having the appearance of long-stemmed, blossoming plant material; the stem is segmented and frequently separates in disklike pieces.

Crust The thin, outermost rocky layer of the earth.

Cryptocrystalline Describing a rock surface or solid rock mass comprised of extremely minute or microscopic crystals.

Cubic Describing an isometric system crystal in the form of a cube.

Cuesta The high exposed end of a sloping plain.

Cueva A cave or grotto.

Delta A deposit (often very large) of sand, gravel, and silt at the mouth of a river.

Dendritic Describing the branching or treelike pattern formed by a foreign mineral within another mineral; also descriptive of some minerals that generate themselves into such branching forms without the involvement of other minerals.

Deposit Mineral or ore concentrations laid down naturally in a specific area.

Diatom Fossilized (often silicified) microscopic marine plant frequently found in sedimentary rocks.

Diggings The excavations made by a prospector or miner; also, the worked mineral site generally.

Dike A flattened vertical body of igneous rock that has intruded through a major mass of country rock without being influenced by country rock layering.

Dip The downward angle of inclination of bedrock from the horizontal.

Divide The crest of land that separates the direction of stream flow in a given area.

Dome A rounded knob of rock, usually large and relatively isolated.

Drift The collection of rock and mineral debris deposited by a glacier at the terminus of its advance or by the runoff waters of that glacier; also, in mining, a horizontal tunnel following a vein or main ore body.

Druse A pocket, vug, or cavity in rock that is lined with crystals of the same minerals of which the rock is comprised (as opposed to amygdule or geode, in which the filling is not related to the country rock).

Drusy   A crust of uniformly tiny crystals filling a cavity or coating a matrix.

Dump   The piles of waste rock that are the residue of mines and prospects.

Eluvial   A deposit formed by the *in situ* disintegration of rocks.

Enhydro   A crystal that contains a fluid-filled cavity, sometimes with a bubble visible through the crystal walls.

Escarpment   An extensive bank of sheer cliffs.

Estuary   A bay that has formed at the mouth of a river where tide flows meet river current.

Evaporite   A chemically originated sedimentary rock or mineral that was formed by the process of evaporation (e.g., halite: rock salt).

Exfoliation   The process of weathered rock sheets splitting away in layers from larger underlying rock masses, leaving the remaining masses essentially rounded.

Extrusives   Igneous rocks that reach the earth's surface while still in a molten state.

Fault   A major fracture in rock along which ordinarily the opposing sides have slid in different directions on the same plane.

Ferruginous   Containing iron.

Fibrous   Containing or consisting of needlelike or threadlike crystal fibers, usually laid in parallel planes (e.g., asbestos).

Fire   The intrinsic brilliance of a gemstone resulting from its characteristic of splitting light rays.

Fissure   An extensive crack or break in a matrix material.

Fissure vein   A band of mineral materials that intruded a fissure while in a molten state.

Flat   A markedly level, usually treeless area.

Float   A general term applied to pieces of rock some distance detached from their outcrop source.

Flow   The hardened residue of a single expulsion of lava.

Flow banding   A structure similar to folding where, in some volcanic rocks, alternating layers of unlike mineral material form and lie adjacent to one another as the result of lava flowing at intervals.

Flower   A form of agate showing a bouquet pattern of variegated colors.

Fluorescence   The emission of brightly colored light by a substance when exposed to ultraviolet rays.

Foliated   Basically describing schist minerals in thin, parallel, sheetlike layers like the pages of a book (e.g., mica).

Fool's gold   Amorphous pyrite.

Formation   A related deposit of rocks formed during the same time period and under generally the same circumstances.

Fortification   A form of agate with an angled striped pattern of generally concentric configuration. (See also *banded.*)

Fossils Plant and animal remains preserved in rock.

Fracture The type of break that occurs in any rock or mineral.

Friable Crumbly material that is broken, pulverized, or separated with facility.

Fumarole A hole or vent through which a volcano issues hot gases.

Gabbro A composition of pyroxene and plagioclase feldspar in a dark granitelike igneous rock.

Gangue The waste material, or material of no commercial value, in a mineral deposit being worked; quartz is often a gangue material.

Geode A hollow mineral shell formed within a cavity of sedimentary rock, often lined with mineral substances including fine crystals of varying types and colors and seldom related to the shell material; easily separable by weathering or artificial methods from the enclosing rock.

Glaciated Subjected to glacial action.

Gneiss A crystalline metamorphic rock that, due to the segregation of its component minerals, has a banded appearance; most frequently comprised of mica, quartz, and a dark mineral such as hornblende, these minerals to some extent laid out in streaks. Gneiss and schist are similar, but gneiss, having less mica in its composition, is stronger.

Gouge A soft material occurring as a layer between country rock and the walls of a vein.

Grain The unit of weight for pearls; equivalent to 0.25 carat.

Gulch A mountain erosion channel or small canyon normally formed by intermittent torrential flows.

Habit The shape in which a crystal characteristic occurs.

Hackle A jagged, noncleavage fracture in a rock or mineral.

Hanging wall The upper portion of country rock bordering a vein or dike; opposite to the foot wall.

Hardness Degree of resistance of a rock or mineral surface to being scratched; the relative 1–10 *Mohs scale* is a measure of hardness, with talc and diamond representing opposite properties of resistance.

Hemimorphic A distinct difference in opposing ends of a crystal due to a lack of transverse planes of symmetry (e.g., zincite and tourmaline).

Hexagonal The crystal system wherein the vertical axis is intersected by three axes at right angles and by a fourth at 60 degrees.

High-grade Used as a verb to connote deliberate selection of the high-value mineral, choicest ore, or gemstone material.

Horizon The particular stratum in which minerals or fossils of similar characteristics appear throughout a sedimentary formation.

Horse A separated mass of country rock enclosed in a vein.

Hyaline   A mineral substance that has a glassy texture.

Hybrid rock   A rock variety formed when one magma combines with another.

Igneous rocks   Rocks formed through crystallization of lava or magma.

Incline   A narrow, oblique shaftlike mine excavation.

Inclusion   The presence of a foreign body (including gas or liquid) enclosed within a crystal. (See also *enhydro.*)

Indurated   Hardened by pressure and/or heat (e.g., indurated clay).

Inorganic   Derived from or pertaining to material of inanimate origin.

Interpenetrant twins   Two or more twin crystals that pass through one another.

Intrusive   Molten material squeezed into cracks or crevices or between already existing rock layers to form an igneous rock that has hardened before reaching the earth's surface.

Iridescence   A rainbow display of colors in a mineral.

Iris   A form of agate that exhibits a distinct spectrum when light passes through it.

Isometric   Crystal system wherein all three axes are of equal length and intersect at right angles.

Joint   A divisive of a plane within a rock mass, occurring in series and intersecting at right angles, with the result of separating the rock into blocks.

Kidney   An oblong, rounded mass of ore or massive gemstone material that in outline form resembles a kidney.

Knob   Any relatively isolated rock hill or mountain.

Lamellar   Comprised of thin, sheetlike layers, scales, or plates.

Laminated rocks   Numerous overlapping layers of sedimentary rock.

Lava   The superheated liquid rock discharge of a volcano.

Ledge   The term used to designate an outcrop of mineral-rich material or to designate a pegmatite outcropping in districts where pegmatites occur.

Lens   A lens-shaped (roughly circular, thick in the middle, and tapering to thin edges) mineral deposit within a body of ore.

Lenticular   Having a flattened, oblong, rounded shape like a lima bean or a lentil.

Level   A horizontal passage (or tunnel or drift) within a mine; such levels are numbered in descending order from the adit.

Lithoclase   A crystal-filled fracture in a rock.

Lithophysae   Distinctive hollow spherulites occurring in rhyolite, obsidian, and similar rocks of glassy character.

Locate   To mark out the boundaries of a claim, establishing possession.

Lode   Essentially the same as a vein.

Loess   Wind-deposited drifts of silt.

Luster   The distinctive characteristic of light reflection from a particular mineral.

Magma   The subcrust molten material (called lava when it reaches the surface) from which igneous rocks derive.

Malleable   Possessing the nonbrittle characteristic that permits a mineral material to be shaped by hammering.

Mammillary   A rounded aggregate of minerals having a breast-like form.

Marl   Calcareous loam or clay that may contain glauconite.

Massive   Having no definite crystal form or structure.

Matrix   The base material (clay, rock, etc.) upon which or within which minerals and/or fossils are perched or embedded.

Mesa   A generally isolated tableland or flat-topped elevation.

Metamorphic rocks   Rocks that have, through action of heat, solutions, and/or gases, become altered in a significant manner.

Meteorite   A celestial body that has fallen to earth.

Mica schist   A basic matrix material whose most conspicuous ingredient is mica.

Mineral deposit   Economically valuable minerals in heavy concentration.

Monoclinic   Crystal system wherein the axes are unequal in length, and two meet at right angles while the third intersects at an oblique angle.

Moraine   An accumulation of debris deposited by the outermost edge of a glacier.

Mother lode   The principal vein or lode; the richest and largest source of a desired mineral.

Nacreous   Exhibiting an opaque pearly luster.

Native   Occurring in a pure or uncombined state, as native copper, native gold, native silver, etc.

Nodule   A roundish lump (generally fist-sized or smaller) of mineral, often formed within a cavity in rock.

Octahedron   Crystal system composed of eight triangular faces.

Oolite   A limestone comprised of numerous small spherules that may be silaceous, calcareous, or ferruginous; these spherules are spherical grains that are less than 2 mm (.08″) in diameter, and the organic or mineral nucleus is successively encased in concentric layerings on various minerals; the term can also be applied to a rock comprised of oolites.

Opalescent   Exhibiting a lustrous milky appearance with a faint but distinct bluish milky cast.

Opaque   Describing a rock or mineral through which light cannot pass.

Opencut   The workings into a mineral deposit via a trench excavation.

Orbicular   Containing a number of solidly encased orbs or spherules with eyelike patterns of alternating mineral material and colors, occurring in metamorphic and igneous rocks. (See also *pisolite.*)

Ore   A deposit of mineral material that is sufficiently rich and of a quantity to be mined with profit.

Ore body   The general continuous mass of material (as distinguished from country rock) in which minerals of value occur.

Organic   Having to do with compounds produced in animals and plants rather than of chemical or otherwise inorganic origin.

Orthorhombic   Describing a crystal system wherein the axes are all of unequal length but wherein all meet at right angles.

Outcrop   The portion of a body of ore, a stratum, or another massive rock form that becomes exposed above the earth's surface.

Overburden   The essentially valueless surface material over a mineral deposit.

Oxidized zone   The portion of an ore body that has been altered through the action of surface waters that carry oxygen, carbon dioxide, etc.

Pan   The process of washing earth, gravel, and sand in a graduated circular pan in search of gold and other heavy materials; also, the center of a basin declivity in the land.

Parting   The occasional tendency of a crystal to separate in an area that is not a true cleavage plane.

Patented claim   A claim legally filed upon and to which permanent rights have been obtained through a conveyance or *patent.*

Pearly   Exhibiting a rich luster similar to that of a pearl. (See also *nacreous.*)

Pegmatite   A body of coarse-grained granite ordinarily occurring in sheetlike masses that have intruded into country rock or formed within the granite itself; most pegmatites exhibit clearly oriented color zones or *units* within which there are characteristic mineral growths.

Perfect   Term for cleavage that is easily instigated along a characteristic plane and that after cleaving leaves a smooth surface.

Perlite   An obsidianlike volcanic rock that has been rendered light and friable due to a multitude of spheroidal and ellipsoid fissures; "Apache tears" are small obsidian nodules that have weathered from perlite masses.

Petrifaction (permineralization)   The process of fossilization; the transitional metamorphic state during which the original cell structure remains but the cells become filled with mineral that itself is often silicified.

Phantos A visible "ghost" crystal that is completely enclosed within another perfectly formed transparent crystal.

Phenocyst A prominent crystal surrounded by considerably smaller grains of similar mineral in porphyritic rock.

Phosphorescence A distinctive form of luminescence that persists after the light stimulus is removed.

Phyllite A lustrous and exceedingly compact schistose rock derived through metamorphic process from clay sediments.

Piedmont The generally shallow-sloping area at the base of mountains.

Pillow basalt Dark basaltic rock flows in which large rounded masses of "pillows" of firm rock were created when the molten material met water; the spaces between these masses filled with friable rock in which occur "pillow" cavities that often contain crystals.

Pica The summit of a mountain.

Pipe A vertical volcanic tube of considerable diameter containing a friable, easily decomposable, cylindrical mass of kimberlite or other igneous rock extending downward to great depth.

Pisolitic Describing a mineral formation of rounded beanlike or pealike masses. (See also *orbicular.*)

Pitch The dip or inclination of a vein or mineral bed at right angles to its strike.

Placer An alluvial deposit often containing gold.

Plateau An extensive flat-topped region of considerable elevation.

Playa A central flat area or depression in a bolson; the dry lakes of California are playas. (See also *pan.*)

Plug A solid lava core located in the neck of an extinct volcano.

Plutonic rocks Rocks formed from molten material that cooled some distance below the earth's surface and generally show distinct grain textures like granite.

Pocket Generally, a small body of ore; in pegmatites pockets are central openings lined with crystals including gem species; term can also be applied to a natural cavity in any rock.

Point A minute diamond weight equivalent to 1/100 carat.

Porphyry An igneous rock comprised of large grains or crystals occurring in fine-grained groundmass; most porphyries are rhyolitic.

Precious Of considerable value. (Accurately, the only gemstones that can be termed precious are diamond, emerald, ruby, sapphire, and opal.)

Prismatic Describing a narrow crystal shape with sides covered with rectangular planes joining in parallel edges having the property of breaking light into its spectrum.

Prospect A "dig," superficial pit, or more extensive working for the purpose of locating the presence of valuable minerals; many mines originate as prospects.

Pseudomorph A mineral crystal having the faces and angles of a different mineral species; the term is also broadly applied to other misleading substitutions such as petrified wood, wherein quartz or opal have become pseudomorphs of wood.

Pumice A bubbly lava that has solidified in so frothy and light a state that it often floats on water.

Punta A headland or point.

Quarry An extensive, deep, steep-sided open excavation worked with machinery and explosives to remove rock, usually for commercial purposes.

Quartzite A conglomerate or sandstone that is composed primarily of coarse quartz grains or pebbles converted to solid rock through metamorphic action.

Radiating Describing a crystal structure in which individual crystals (often acicular) fan outward in a starburst pattern from the center. (See also *stellate.*)

Ravine A natural erosion trench or valley with steep sides; generally speaking, smaller than a canyon but larger than a gully.

Reniform A kidney-shaped mineral form.

Replacement The process through which one mineral assumes the place and (often) the shape of another. (See also *pseudomorph.*)

Reticulated Having a lattice or network formation.

Rhyolite Volcanic rock species that is the equivalent of granite and contains tiny orthoclase and quartz crystals, along with biotite and hornblende, or a pyroxene; most rhyolites are light in color.

Rough Uncut, unpolished gemstone material.

Rutilated Crystal with inclusions of rutile needles (titanium oxide) laid out most often in a helter-skelter jackstraw pattern rather than lying parallel.

Sagenitic Describing a crystal that contains smaller needlelike crystals of a foreign mineral.

Schist Foliated (book-sheet) metamorphic rocks characterized by long, drawn-out streaks and containing considerable mica, causing it to be brittle as well as readily separable into larger sheets along the layers.

Seam A thin layer, stratum, or vein of mineral.

Sectile Describing a mineral soft enough and malleable enough that it can readily be cut into shavings.

Sedimentary rocks Rocks formed in layers through the accumulation and solidification of animal, mineral, or vegetable matter; water, wind, and glaciers are the principal means by which the loose sediments are transported and deposited.

Semiprecious Nonprecious gemstone minerals that can be cut, polished, and sometimes worn as gems (including jasper, agate, chalcedony, etc.).

Shaft A deep and narrow mining excavation sunk vertically.

Shale A sedimentary rock formed by mud or silt under great pressure.

Shear zone A rock mass in which a zone of shearing has occurred and this shear zone has become filled with crushed and brecciated rock.

Sheet A usually extensive bed of eruptive rock that has intruded between layers of preexisting rock; also, a similar body of volcanic rock that has outpoured upon the earth's surface in a layer of relatively uniform thickness.

Shingle Generally uniformly sized pebbles of waterworn beach gravel.

Siliceous (silicified) Describing a mineral in which silica or quartz is present; may also be applied to any animal, vegetable or mineral substance that has been replaced by chalcedony or opal.

Silky Imbued with the luster and tactile sensation of silk.

Sill A broad, flat ribbon or thin sheet of igneous rock that has intruded parallel to the bedding of other rocks and lies in a more or less horizontal plane.

Sinkhole A very steep-walled craterlike depression; usually occurs in limestone regions due to the roof collapse of a subterranean cavern.

Slate A rock that is similar in character to shale but forms in a more highly compressed state and is more easily split along the bedding planes; small scales of mica are often present, giving the slate a distinctive silvery luster.

Sluice (sluice box) A narrow wooden inclined trough with a bottom ribbed with properly spaced bars called *riffles* over which gravel is poured and washed down in order to extract the heavier minerals at the riffles.

Spall To break off slabs or pieces of rock from a much larger rock mass.

Specific gravity The comparison of the weight of a mineral substance with that of the water it displaces upon being immersed.

Spherulites Distinctly rounded aggregates or rosettes of very fine, delicate, needlelike crystals radiating from common centers; orbicular jaspers tend to exhibit some spherulitic structure.

Splendant Brilliantly reflective.

Spoil Debris or waste materials removed from a mine, prospect, or other excavation.

Stalactite A tapered column of mineral matter depending from the ceiling of a cavern; its growth is determined by mineral particles contained in dripping water.

Stalagmite The opposite of a stalactite; a tapered pillar rising from the floor instead of depending from the ceiling; its growth is often determined by the dripping of water from a stalactite directly above.

Star   An astirated gem, or one that has the quality of reflecting a four- or six-rayed star from embedded fibers (e.g., star ruby, star sapphire).

Stellate   Having a radiate crystalline structure forming starlike patterns.

Stock (stockwork)   A mass of country rock in which many small veins have developed into a complicated network.

Stope   The series of "steps" above or below a mine level where ore has been extracted by excavating an inclined ore body.

Stratum   A layer of rock, not necessarily horizontal but of similar constituent character throughout.

Streak   The colored mark that is left behind when a mineral is rubbed on an unglazed white porcelain tile; a method of mineral identification.

Striation   A minute groove or channel on a crystal face.

Strike   The horizontal direction or trace of an outcrop.

Stringer   An especially narrow vein (or veinlet) that is generally irregular in its thickness and follows an erratic course.

Structure   The general visible character of a rock specimen or outcrop.

Symmetry   The regular, balanced arrangement of the features, forms, and properties of any given crystal relative to its axes.

Tabular   Describing broad, flattened crystals of tablet shape (e.g., Wulfenite).

Talus   A sloping heap of (usually) naturally fallen debris at the base of a mountain, cliff, or promontory.

Tenacity   The ability of a mineral to resist breaking or separation.

Termination of a crystal   The group of planes completely enclosing the crystal's end; a doubly terminated crystal has both ends covered with faces and thus ordinarily has its entire surface comprised of crystal faces.

Tetragonal   Describing a crystal system wherein only two axes are equal in length and all three axes intersect at right angles.

Thunderegg   A nodule (usually fist-sized or smaller) having a colored agate core and a surface structure of rhyolite or jasper.

Translucent   Describing a crystal with the ability to pass light but not the image of an object.

Transparent   Describing a crystal with the ability to pass light so clearly that objects can be observed through it.

Traprock   Basalt and numerous other dark, fine-grained igneous rocks.

Triclinic   Describing a crystal system wherein all three axes are unequal in length and in which they meet at oblique angles.

Trilobite   Ancient Paleozoic fossil crustacean that bears a resemblance to the modern sow bug or wood louse but was much larger.

Trisoctahedron   Crystal system formed of 24 faces so arranged that three take the place of each face of a regular octahedron.

Tuff   A well-cemented but easily breakable volcanic ash.

Twin   A compound crystal that is comprised of two or more crystals (or parts of crystals) in reversed positions with respect to each other; also, a specimen that consists of two or more single crystals of the same mineral intergrown in a definite systematic arrangement.

Vein   An irregular twisting mineral deposit of considerable length that is very thin in proportion to its length and breadth; veins are normally formed as a result of hydrothermal activity.

Vesicle   A small cavity found in glassy and volcanic igneous rocks; such vesicular cavities are formed through the expansion of gases.

Vitreous   Describing a rock or mineral that is like glass in physical properties, in appearance, or both.

Vug   A cavity, usually small, in rock; it is normally lined with crystals firmly fastened to the walls or scattered on the bottom that are related to the minerals found in the country rock; vugs are believed to be caused by extensive shrinkage of the enclosing rock.

Wash   The loose surface debris that is found in the bottoms of canyons and intermittent streams; a wash is also the generally broad and shallow depression in which such debris occurs.

Xenolith   Country rock enclosed in magma.

INDEX

KEY TO PRINCIPAL ABBREVIATIONS AND MINERAL LOCATION NUMBERS

Principal Abbreviations

aka also known as
BLM Bureau of Land Management
Byp bypass route
CF collecting fee
cm centimeters
CR county road
ct carat
I interstate highway
lb pounds
MM mile marker
mm millimeters
oz ounces
RR railroad
r/w right-of-way
SR state highway
US U.S. highway
var. variety
″ inches
′ feet

Mineral Location Numbers

1. On the surface and/or in the soil
2. In mine and/or in mine dump
3. In gravel deposit
4. In natural or man-made exposure and/or cut
5. In matrix or *in situ*

6. In soil (but usually *not* on surface)
7. In stream bed, lake bed, coastal shallows
8. In talus and/or natural debris
9. In vugs and/or pockets and/or cavities
10. In gravel pit
11. In sand pit
12. In quarry
13. In washes, ravines, gullies, draws, ditches
14. Weathered from matrix
15. In volcanic tuff or pillow basalt
16. In pegmatite dike
17. In gold-bearing sands and/or gravels
18. In/on gravel bar
19. In vein or seam
20. Loose on shoreline
21. In prospect or placer
22. In sediment deposits or sedimentary strata
23. In alluvial deposits
24. In exposed pegmatite dike
25. In clay
26. In sand
27. In schist
28. In shale and/or slate
29. In outcropping (natural or exposed by man)
30. In spoil deposits
31. In ash deposits
32. In cave and/or cave vicinity
33. In marl
34. In gneiss
35. In clay pit
36. In/on sand bar
37. In/on fill deposit
38. Occurring as drusy or crust
39. In natural bed or deposit
40. In conglomerates
41. In concretions
42. In/on ledges